ECI Pricing Standard
for Piping Works

ECI Benelux Subcontracts Taskforce

Thomas Telford

Published by Thomas Telford Publishing, Thomas Telford Ltd, 1 Heron Quay, London E14 4JD.
URL: http://www.thomastelford.com

Distributors for Thomas Telford books are
USA: ASCE Press, 1801 Alexander Bell Drive, Reston, VA 20191-4400, USA
Japan: Maruzen Co. Ltd, Book Department, 3–10 Nihonbashi 2-chome, Chuo-ku, Tokyo 103
Australia: DA Books and Journals, 648 Whitehorse Road, Mitcham 3132, Victoria

First published 2002

A catalogue record for this book is available from the British Library

ISBN: 0 7277 3120 3

© The European Construction Institute, 2002

Typeset by Kneath Associates, Swansea

Printed and bound in Great Britain

Contents

Foreword

Why would you use this ECI Piping Pricing System, when you already have your own system in place?

A perfectly sensible question. However, the Piping Workgroup feels that there are a number of good reasons for implementing this ECI Piping Pricing System, since existing systems do not offer you all of the following:

- separation of direct costs from indirect costs
- possibility to separate fully piping prefabrication cost from erection cost (i.e. cater for separate contracts)
- measurement corresponding to basic physical operations and related to man-hour content per operation
- fair representation of effects from quantity variations and from complexity variations (recognises piping complexity, i.e. weld/length ratio)
- possibility of linking Material Take-off (MTO) and isometrics Bill of Material (BOM) data for electronic pricing
- allow basic price agreement at a very early stage, e.g. for pre-selection of potential partners
- allow logical links with estimating data collection systems, progress measurement data collection systems and historical data collection systems
- provide for transparent change administration.

The ECI Piping Pricing System offers all of the above features. This document explains how it works.

The system includes a spreadsheet with multipliers on a CD-ROM. It is not a ready-made software package, but it is simple enough to integrate the system into any company's existing software.

This Piping Pricing System has been jointly developed by a group of professional and committed Engineering and Construction Contractors. Such a broad basis is a guarantee of wide acceptance by others in the business.

The ECI are grateful to the Subcontracts Taskforce and the Piping Pricing System Workgroup for their efforts in developing this ECI Piping Pricing System.

This document is dedicated to the memory of Michel Buidin and François Dalla Vecchia. They will be remembered for their expert contribution and good fellowship.

Cees Zwinkels and *Gerard Bakker*,
Chairmen, ECI Benelux Subcontracts Taskforce

1. Introduction

1.1 General

Developments in the petrochemical and chemical business over recent years have increased the demand for innovation, efficiency and flexibility from Owners, Engineering Contractors and Subcontractors (or Construction Contractors) alike.

The Benelux Regional Unit of ECI has taken the initiative to form a Subcontracting Taskforce investigating options to improve subcontract processes.

The members of this Taskforce are people with practical day-to-day experience with (sub)contract and construction matters. Combining the knowledge, expertise and experience available within the members of the Unit has provided the platform for the development of improvements for the benefit of all.

One of the Taskforce's objectives is to develop a new standardized/uniform pricing system for Piping Subcontracts, which will be widely accepted in the construction industry. The numerous, very different types of systems in use to date are creating significant problems for Contractors in properly assessing and pricing work. This often leads to an unrealistic level of contingencies in proposals. Misunderstandings, unclear items and/or quantity variations cause disputes and claim situations after award. The new system aims to eliminate, to a large extent, the drawbacks experienced in current systems.

For ease of reference, we will use the term 'Employer' for Engineering Contractor (or Owner as the case may be) and 'Contractor' for Subcontractor or Construction Contractor.

Subcontracting Taskforce

Taskforce Objectives....

"Improve the subcontracting process on industrial projects"

- Reduce bidding time
- Better identify and manage risk
- Avoid claims by fair cost assessment of
  - scope of work
  - variations

The primary goals for the new system are:

(a) **establish a less complex, market-friendly system, representing actual costs per operation**

the cost of a single physical operation can be defined better than the combined cost of a number of diverse operations

(b) **take risk away from those parties that cannot control such risks**

reduce the number of assumptions a contractor has to make in his bid by separating prices for prefabrication and erection work, recognition of complexity, and providing a transparent system which will absorb quantity variations without requiring re-negotiations

(c) **establish a method for electronic pricing of piping work**

(d) **ensure transparency in measurement of progress and productivity.**

1.2 Phases of development

**A specification or 'shopping list' was established, listing
requirements that the ECI Pricing System would have to
meet. The following parameters were considered as being
vital:**

- recognize piping complexity (weld-length ratio)
- segregate small bore from large bore piping
- relate to man-hour content per operation and in total
- separate prefabrication from erection activities for large bore
  piping
- have a direct connection with engineering MTO data
- allow basic price agreement in very early stage (e.g. for
  selection of alliance partner)
- separate direct costs from indirect costs and differentiate
  between operations (e.g. Non Destructive Examination (NDE),
  painting, etc.)
- allow a logical link with progress measurement systems and
  planning/work preparation
- support estimating systems and provide logical collection of
  historical data
- allow practicable administration
- allow transparent change administration
- suitable for Electronic Data Transfer via e-mail, CD-ROM or
  diskette
- suitable for global application.

Various systems in use to date were presented, commented on
and evaluated. These systems were analyzed in detail and tested
against the above criteria.

It was observed that one recently developed system, the
Raytheon 'New Pay Item System' came nearest to meeting most
requirements and was suitable for further development into the
envisaged ECI Piping Pricing System.

Team members of the Piping Pricing System Workgroup that were involved in the setting-up of the system were:

Gerard Bakker	Fluor Daniel (Workgroup chairman)
Michel Buidin	Fabricom
François Dalla Vecchia	Ponticelli
Willem Graulus	Fabricom
Rob van Hoeve	Stork ICM
Tony Kohlen	ABB Lummus Global
Teun Noordam	Stork ICM
Michel Sauvage	Ponticelli
Michel Stoelinga	Raytheon Engineers & Constructors
Cees Zwinkels	Kvaerner Process (Previously Raytheon, former workgroup chairman)

Piping Unit Rate System

Main System Characteristics....

- **Transparent representation of operations**
  - Clear description of Units of Work
  - Separation of unrelated cost items
  - Better representation of actual costs
- **Reduce risks contractors can't control**
  - Reducing assumptions during bidding
  - Fair representation of effects from quantity variations and complexity
- **Fixed multipliers reduce bidding time and final re-measurement time**

1.3 The ECI Piping Pricing System

The system is based on measurement corresponding to physical operations and defined interrelations between variables within an operation. Indirect costs are separated from direct costs.

Operations (Units of Work) recognized are:

- handling
- jointing
- bending
- transportation
- hydrotesting, NDE, Post-Weld Heat Treatment (PWHT)
- tapering, cutting, beveling
- painting
- supporting.

In this system each **activity** or **operation** performed will be measured. It recognizes separate measurement for prefabrication and erection. Interrelations between variables (size, rating, schedule, etc.) have been defined in tables. These interrelations are reflected in multiplier factors. For instance, a **reference pipe** has been defined (for large bore) which is **6 inches** (in.), schedule **STD**, **Carbon Steel**. The Contractor will quote a **Base Unit Rate** for this 'reference' item and all other items are scaled from this unit rate by applying the relevant defined multipliers.

1.4 Example

The Unit Rate for an Erection Butt-Weld on Stainless Steel, diameter 4 in., Schedule 10S, shall be calculated as follows:

Base Unit Rate: Weld/each (Assumption)	Multiplier: Material type SS	Multiplier: Pipe size 4 in.	Multiplier: Wall thickness Schedule 10S	Calculated Unit Rate
100.00	x 1.6	x 0.8	x 0.8	= 102.40

The apparent relation between the labour content and the cost of an operation enables expedient verification of estimated man-hour content as well as close monitoring of progress.

*Simplified*Multiplier Table (2) ECI

MATL TYPE		HANDLING	WELD JOINT	HANDLING INLINE ITEMS	FLANGED JOINT		PIPE DIAMETER		HANDLING	WELD JOINT	HANDLING INLINE ITEMS	FLANGED JOINT
	CARBON STEEL	1,0	1,0	1,0			2"	0,70	0,60	0,55	0,65
	KILLED CARBON STEEL	1,0	1,0	1,0			3"	0,80	0,70	0,70	0,70
	STAINLESS STEEL	1,1	1,6	1,0			4"	0,90	0,80	0,80	0,80
	ALLOY STEEL	1,0	2,0	1,0			6"	1,00	1,00	1,00	1,00
							8"	1,10	1,40	1,25	1,25
WALL SCHEDULE	5S, 10S	0,8	0,8				10"	1,30	1,80	1,50	1,50
	10, 20, 30, 40, STD	1,0	1,0				12"	1,50	2,10	1,75	1,75
	60, 80, XS	1,2	1,3				14"	1,75	2,30	2,00	2,00
	100, 120, 140, 160, XXS	1,6	2,0				16"	2,00	2,60	2,25	2,25
	25.5 -50mm WT	2,2	3,0				18"	2,25	3,00	2,50	2,50
							20"	2,50	3,30	2,75	2,75
FLANGE RATING	150 lbs.			1,0	1,0		24"	3,00	3,90	3,25	3,25
	300 lbs.			1,15	1,15		26"	3,20	4,20	3,50	3,50
	600 lbs.			1,3	1,3		30"	3,60	4,90	4,00	4,00
	900 lbs.			1,5	1,5		32"	3,80	5,30	4,30	4,25
	1500 - 2500 lbs.			2,0	2,0		36"	4,20	6,10	5,00	4,75

1.5 Application

The system, as presented, is intended for use on Unit Rate type Piping Contracts. In the preparation, the following activities have *not* been considered for inclusion:

- detailed engineering
- supply of piping materials
- activities such as hydraulic bolt tensioning, Positive Material Identification (PMI) and chemical cleaning
- work of other disciplines, e.g. mechanical, scaffolding, steelwork, instrumentation and insulation.

1.6 Sources used

For the determination of interrelations and multiplier factors, a number of sources of information were consulted, and figures compared and evaluated to establish a commonly agreed basis:

- Linde system
- Pay Item System, Raytheon Engineers & Constructors
- *Estimators Piping Manhour Handbook*, Page & Nation, USA
- SNCT (Syndicat National de la Chaudronnerie et de la Tuyauterie) of France
- Ponticelli Frères database
- Fabricom FEMS system

The System's "Multipliers"

Basis of Multipliers....

- Statistics of participating ECI members
- Page & Nation (USA)
- SNCT (France)
- Linde (Germany)
- And......

subject to,......
and result of,......
extensive discussions!

2. Description of the system

2.1 General

2.1.1 Basis of the ECI Piping Pricing System

The heart of the New Pricing System is formed by the **Multiplier Tables**: one table for Prefabrication activities and one for Erection.

All relevant factors influencing Operations in a (unit rate based) Piping Contract are shown in table format. For each Operation, **fixed** multiplier factors are given, representing their relation to the 'Reference Operation'.

Only Base Unit Rates for the 'Reference Operations' are required, other unit rates are automatically scaled off by applying the relevant multipliers.

The multiplier factors have been established and verified as a true representation of the related complexity and shall **not** be altered. One of the main virtues of the system is that it may save considerable time in preparation of a bid due to the standardized and accepted nature. Modifications to, or discussions about, multipliers would jeopardize this.

It must be noted that some piping related activities do not appear in the Multiplier Tables because they are specific and no Multiplier factors apply (e.g. safety valve testing). For each of these, relevant Pricing Tables can be prepared and applied.

Contractors shall take the characteristic nature of the work of the project into account on the basis of all relevant information provided in the bid package.

2.1.2 Units of Work

In this system, the work scope is broken down into operations for which (unit) prices are agreed. These activities, which are all (re-)measurable operations, are referred to as *Units of Work*.

These Units of Work may also include items that will **not** be measured or paid for separately. Such items are referred to as 'non-measured items'.

The prices for the Units of Work are deemed to include the associated non-measured items.

2.1.3 Units of Work Numbering System

Each user may assign his own coding system for numbering the different Units Of Work, suitable for intelligent coupling to the material take-off system employed, for estimating and progress control purposes.

2.1.4 Measurement

During the bid phase, the Employer will provide a Form of Tender which includes the various applicable Pricing Tables. Units of Work quantities in these Pricing Tables are usually provided from estimated Material Take-Offs.

During the execution phase, actual Units of Work quantities are best calculated and measured on the basis of Released for Construction drawings. The method of measurement (referred to as Measurement) indicates how the Unit of Work quantities may be measured.

For progress measurement, a Contractor shall usually submit their determination of Units of Work performed in accordance with the provisions of the Contract.

2.1.5 Description Work Breakdown Structure

The following breakdown structure has been used for reference.

2.1.5.1 Prefabrication

(a) Transportation of piping materials — agree suitable means of compensation (see sample set-up).

(b) Shop prefabrication and painting activities, as included in Units of Work:

- Handling
- Joints
- Testing and Heat Treatment
- Supports

■ Painting

■ Modification work.

(c) Labour and Equipment rates — agree suitable means of compensation.

2.1.5.2 Erection

'Indirect Costs' shall be clearly separated from 'Direct Costs'.

'Indirect Costs' cover those items of expense that do not vary proportionally with the quantities of work.

'Direct Costs' cover those items of expense that vary with the quantities of work.

(a) Indirect costs — compensation for the following items of expense shall be excluded from the unit rates and shall be priced separately as 'Indirect Costs'.

■ *Site and/or Camp Facilities*: Mobilization, Demobilization, Operation and Maintenance Costs, including man-hours spent in relation to these items and all running costs such as electricity, phone, fax, stationery, computers, etc.

■ *Site Staff, Management and Supervision*: Management, Supervision above level of working foreman, Administration, Drawing and Design, Planning, Preparation and Follow-up of work, including Quality Assurance/Quality Control (QA/QC), Safety, etc. and relative costs for mobilization/demobilization according to a list of functions and expected agreed assignment periods.

■ *Major Construction Equipment*: limited to Cranes, Rigging Equipment, On-site Transportation Equipment for Personnel and Material, Power Generators, Air Compressors, including their Operators if applicable and all relative costs for mobilization/demobilization, consumables, maintenance, etc. according to a list of items and expected agreed assignment periods.

■ *General Services*: cost of Personnel assigned to Tools and Equipment Warehousing and Maintenance, except for those in relation to the above Major Construction Equipment.

■ *Overheads and Profits*: limited to the part in relation to items of expense that form part of the 'Indirect Costs'.

Prices and/or Rates covering 'Indirect Costs' remain only valid insofar as 'Quantities', 'Schedule' and the like do not change beyond a mutually agreed percentage or period.

(b) Direct Costs to cover Piping Erection and Painting Activities — items of expense which are not covered in the 'Indirect Costs' shall be covered by the unit rates that are to be established for:

- Handling
- Joints
- Bends
- Testing and Heat Treatment
- Supports
- Painting
- Modification work.

(c) Labor and Equipment rates — agree suitable means of compensation.

2.1.6 Guidelines/recommendations

- Detailed scope description: split of work between Employer and Contractor may vary from project to project — this should have no effect on the application of the system.
- Project specific extensions to the system are possible: e.g. for piping material types not covered, a multiplier may be added as the need arises — however, any extensions shall be clearly identified as such.
- Small bore piping is **not** split in 'Prefabrication' and 'Erection'. This decision is left to the Contractor to seek the most efficient solution in each instance. It shall therefore be considered that the unit rate quoted under 'Erection' is the aggregate of a mix of shop and field work as estimated by the Contractor.
- In case of a 'prefabrication only' Contract, any small bore involved shall be calculated according to the Erection Multiplier Table.
- Wall schedule multipliers quoted refer to ANSI standard wall-thickness. For DIN pipe, the wall-thickness in millimeters shall determine the relevant factor on the basis of the nearest ANSI equivalent thickness.

2.1.7 Exceptions

■ Multiplier Tables do not cover a wall-thickness > 2 in. (50 mm).

■ Material types, other than those listed in the Multiplier Tables are to be agreed as required.

■ Multiplier Tables do not cover a pipe > 48 in. (DN 1200).

■ Multiplier Tables do not cover Flange Ratings > 2500 lbs.

■ Multiplier Tables do not cover the bending of a large bore/long radius pipe.

■ Although steam tracing is not covered by the system, supply and return lines can be regarded as small bore piping.

■ Demolition and Tie-in work are not covered in this system.

It is recommended that handling of these excepted items is agreed as required.

2.1.8 Configuration of the system

This handbook does **not** provide a ready-made software package to load and implement into the system of any organization: there are simply too many different hardware and software options in use on the market to restrict this system to any particular configuration. It is however simple enough for each user to start from what is available and standard within his or her own organization and develop the software requirements for the system to suit the particular situation. The system can be set up to run on any commercially available database, spreadsheet or similar type of application.

A parts library will have to be set up to link MTO (Materials Take-Off) data, generated by computer plant design tools such as PDS, PDMS, or similar, with the Pricing Tables and related multipliers therein. As soon as isometrics are produced, the isometric material lists can be linked to the relevant pricing data to generate cost per isometric.

2.2 Pricing Structure diagrams

The Pricing Structures for the Piping (including painting) Work is split into **Prefabrication** and **Erection**, thus providing the possibility of having two separate Contracts.

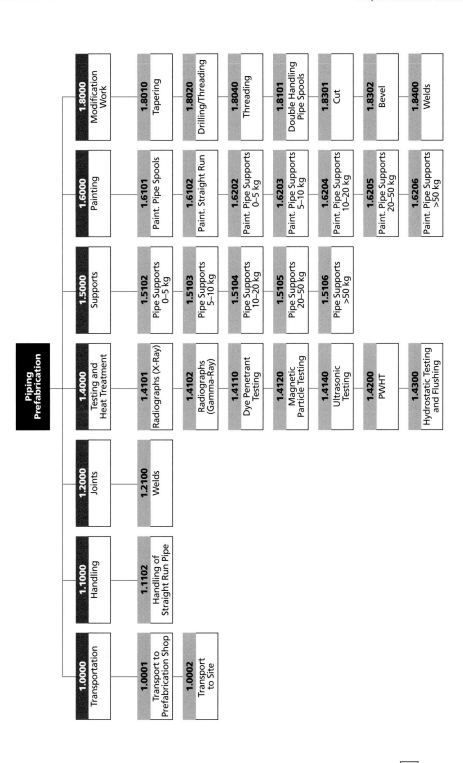

Piping Prefabrication

- **1.0000** Transportation
  - **1.0001** Transport to Prefabrication Shop
  - **1.0002** Transport to Site
- **1.1000** Handling
  - **1.1102** Handling of Straight Run Pipe
- **1.2000** Joints
  - **1.2100** Welds
- **1.4000** Testing and Heat Treatment
  - **1.4101** Radiographs (X-Ray)
  - **1.4102** Radiographs (Gamma-Ray)
  - **1.4110** Dye Penetrant Testing
  - **1.4120** Magnetic Particle Testing
  - **1.4140** Ultrasonic Testing
  - **1.4200** PWHT
  - **1.4300** Hydrostatic Testing and Flushing
- **1.5000** Supports
  - **1.5102** Pipe Supports 0–5 kg
  - **1.5103** Pipe Supports 5–10 kg
  - **1.5104** Pipe Supports 10–20 kg
  - **1.5105** Pipe Supports 20–50 kg
  - **1.5106** Pipe Supports >50 kg
- **1.6000** Painting
  - **1.6101** Paint. Pipe Spools
  - **1.6102** Paint. Straight Run
  - **1.6202** Paint. Pipe Supports 0–5 kg
  - **1.6203** Paint. Pipe Supports 5–10 kg
  - **1.6204** Paint. Pipe Supports 10–20 kg
  - **1.6205** Paint. Pipe Supports 20–50 kg
  - **1.6206** Paint. Pipe Supports >50 kg
- **1.8000** Modification Work
  - **1.8010** Tapering
  - **1.8020** Drilling/Threading
  - **1.8040** Threading
  - **1.8101** Double Handling Pipe Spools
  - **1.8301** Cut
  - **1.8302** Bevel
  - **1.8400** Welds

Piping Erection

2.1000 Handling
- **2.1101** Handling of Pipe Spools
- **2.1102** Handling of Straight Run Pipe
- **2.1200** Handling of Non-welded In-line Items
- **2.1300** Handling of Safety/Relief Valves

2.2000 Joints
- **2.2100** Field Welds
- **2.2200** Flanged Joints
- **2.2250** Hydraulic Bolt Tens. of Flanged Joints
- **2.2300** Screwed Joints

2.3000 Bends

2.4000 Testing and Heat Treatment
- **2.4101** Radiographs (X-Ray)
- **2.4102** Radiographs (Gamma-Ray)
- **2.4110** Dye Penetrant Testing
- **2.4120** Magnetic Particle Testing
- **2.4140** Ultrasonic Testing
- **2.4200** PWHT
- **2.4300** Hydrostatic Testing

2.5000 Supports
- **2.5101** Small Bore Pipe Supports
- **2.5102** Pipe Supports 0–5 kg
- **2.5103** Pipe Supports 5–10 kg
- **2.5104** Pipe Supports 10–20 kg
- **2.5105** Pipe Supports 20–50 kg
- **2.5106** Pipe Supports >50 kg
- **2.5301** Spring Hangers 0–20 kg
- **2.5302** Spring Hangers 20–50 kg
- **2.5303** Spring Hangers >50 kg

2.6000 Painting
- **2.6100** Cleaning & Painting of Welds
- **2.6101** Final Painting of Installed Pipe
- **2.6201** Field Painting Small Bore Pipe Supports
- **2.6202** Paint. Pipe Support 0–5 kg
- **2.6203** Paint. Pipe Support 5–10 kg
- **2.6204** Paint. Pipe Support 10–20 kg
- **2.6205** Paint. Pipe Support 20–50 kg
- **2.6206** Paint. Pipe Support >50 kg

2.8000 Modification Work
- **2.8010** Tapering
- **2.8020** Drilling/Threading
- **2.8040** Threading
- **2.8101** Double Handling Pipe Spools
- **2.8102** Double Handling Straight Run Pipe
- **2.8103** Double Handling In-line Items
- **2.8301** Cut
- **2.8302** Bevel
- **2.8400** Field Weld
- **2.8500** Flanged Joint
- **2.8550** Breaking Flanged Joint

2.3 Multiplier reference points

Unit Rates are calculated from Base Unit Rates which are multiplied with the applicable Factors for Material Type, Diameter, Wall Schedule, Flange Rating and Special Items/Weld Type, as set in Multiplier Tables MT-1 and MT-2.

The Multipliers are fixed and firm, and shall not be changed for changes in scope, quantity, complexity or any other reason.

The reference points (Base Multiplier = 1.0) for the Base Unit Rates are as follows:

Material Type	**Carbon steel**
Pipe/Fitting Size	**For large bore: diameter 6 in. (or DN150)** **For small bore: diameter 3/4 in. (or DN20)**
Wall Schedule	**For large bore: standard** **For small bore: schedule 80**
Flange Rating	**150 lbs**

2.4 Description for Piping Prefabrication Multipliers (Multiplier Table MT-1)

2.4.1 Multipliers for Material Type

The Base Unit Rates for Handling, Welds, Post-Weld Heat Treatment (PWHT), Tapering, Cuts and Bevels are based on:

Carbon Steel

(for which the Multiplier is set as *1.0*)

The Multipliers for all other Material types, as listed in Multiplier Table MT-1, shall be applied to the above mentioned Base Unit Rates to cover variations to the extent of the Work involved with the applicable Material types.

2.4.2 Multipliers for Pipe/Fitting Size

The Base Unit Rates for Handling, Welds, NDE, PWHT, Tapering, Cuts and Bevels, and for Painting are based on:

For Large Bore Piping: Outside Diameter 6 in. (or DN150)
For Small Bore Piping: Outside Diameter 3/4 in. (or DN20)

(for which the Multipliers are set as *1.0*)

The Multipliers for Pipe/Fitting Size for all other outside diameters, as listed in Multiplier Table MT-1, shall be applied to the above mentioned Base Unit Rates to cover variations to the extent of the Work involved with the applicable outside diameters.

2.4.3 Multipliers for Pipe Wall Schedule

The Base Unit Rates for Handling, Welds, NDE, PWHT, Tapering, Cuts and Bevels are based on:

For Large Bore Piping: Wall Schedule STANDARD (STD)
For Small Bore Piping: Schedule 80
(for which the Multiplier is set as *1.0*)

The Multipliers for Pipe Wall Schedule for all other wall schedules, as listed in Multiplier Table MT-1, shall be applied to the above mentioned Base Unit Rates to cover variations to the extent of the Work involved with the applicable Wall Schedules. For DIN pipe, refer to Section 2.1.6, Guidelines/recommendations.

For any wall thickness between 1 in. (25.5 mm) and 2 in. (50 mm), the multiplier for '25.5–50 mm wall-thickness' shall prevail.

2.4.4 Multipliers for Weld Types/Special Items

The Base Unit Rates for Weld Types are based on:

Butt Weld/Fillet Weld
(for which the Multiplier is set as *1.0*)

The Multipliers for all other Weld Types, as listed in Multiplier Table MT-1, shall be applied to the above mentioned Base Unit Rate to cover variations to the extent of the Work involved with the applicable Weld Types. Note that for Welds on slip-on flanges, Multiplier **1.0** shall also be used.

2.4.4.1 Multiplier for Welded Valves

The Multiplier for large bore Welded Valves applies to the Base Unit Rate for each Weld and covers the extra work involved in the dismantling, handling and mounting of valve internals.

2.4.4.2 Multiplier for Socket Welds

The Multiplier for welding Socket Welds applies to the Base Unit Rate for Welds and covers the reduced work involved with welding a Socket Weld as against a Butt Weld.

2.4.4.3 Multiplier for Seal Welds

The Multiplier for welding Seal Welds applies to the Base Unit Rate for Welds and covers the reduced work involved with welding a Seal Weld as against a Butt Weld.

2.4.4.4 Multiplier for Angle Welds/Dummy Legs

The Multiplier for welding Angle Welds and/or Dummy Legs applies to the Base Unit Rate for Welds and covers the extra work involved with welding an Angle Weld for a mitre-bend, or for welding a Dummy Leg to a pipe.

2.4.4.5 Multiplier for Branch Welds (90°)

The Multiplier for Branch Welds (including nipples, half-couplings and bosses) applies to the Base Unit Rate for Welds of the branch size and covers the extra work involved with drilling/cutting a hole in, and welding a branch to, the main pipe under a straight angle of 90°.

2.4.4.6 Multiplier for Branch Welds (angle not 90°)

The Multiplier for Branch Welds applies to the Base Unit Rate for Welds of the branch size and covers the extra work involved with drilling/cutting a hole in, and welding a branch to, the main pipe under an angle that is not 90°.

2.4.4.7 Multiplier for Reinforced Branch Welds (90°)

The Multiplier for Reinforced Branch Welds applies to the Base Unit Rate for Welds of the branch size. It covers the extra work involved with drilling/cutting a hole in, and welding of a branch to, the main pipe under a straight angle of 90°, as well as fabrication from pipe material, and handling and welding a Branch Reinforcement Pad.

2.4.4.8 Multiplier for Reinforced Branch Welds (angle not 90°)

The Multiplier for Reinforced Branch Welds applies to the Base Unit Rate for Welds of the branch size. It covers the extra work involved with drilling/cutting a hole in, and welding a branch to, the main pipe under an angle that is not 90°, as well as fabrication from pipe material, handling and welding a Branch Reinforcement Pad.

2.4.4.9 Multiplier for Welding Outlets (90°)

The Multiplier for Welding Outlets (Weldolets, Flangolets, Sockolets, Nippolets, Threadolets) applies to the Base Unit Rate for Welds of the Outlet size. It covers the extra work involved with drilling/cutting a hole in, and welding an Outlet to, the main pipe under a straight angle that is 90°.

2.4.4.10 Multiplier for Welding Outlets (angle not 90°)

The Multiplier for Welding Outlets (Latrolets, Elbolets) applies to the Base Unit Rate for Welds of the Outlet size and covers the extra work involved with welding an Outlet to the main pipe under an angle that is not 90°.

ECI Multiplier Table MT-1: PIPING PREFABRICATION UNIT RATE MULTIPLIER TABLE

			HANDLING R.L. + MODS	WELD JOINT	NDE, X + GAMMA RAY	NDE, US/DP/MP	PWHT	HYDRO TEST	TAPERING	CUTS/ BEVELS	PAINTING
MATERIAL TYPE	**CARBON STEEL**		1.0	1.0	1.0	1.0	1.0	1.0	1.0	1.0	
	KILLED CARBON STEEL		1.0	1.0	1.0	1.0	1.0	1.0	1.0	1.0	
	STAINLESS STEEL		1.1	1.4	1.0	1.0		1.0	1.2	1.2	
	ALLOY STEEL		1.0	1.4	1.0	1.0	1.2	1.0	1.2	1.2	
PIPE/FITTING SIZE LARGE BORE	2 in.	DN50	0.70	0.60	0.80	0.80	1.00	0.70	0.60	0.60	0.40
	3 in.	DN80	0.80	0.70	0.85	0.85	1.00	0.80	0.70	0.70	0.55
	4 in.	DN100	0.90	0.80	0.90	0.90	1.00	0.90	0.80	0.80	0.70
	6 in.	**DN150**	**1.00**	**1.00**	**1.00**	**1.00**	**1.00**	**1.00**	**1.00**	**1.00**	**1.00**
	8 in.	DN200	1.10	1.40	1.10	1.20	1.00	1.10	1.40	1.40	1.30
	10 in.	DN250	1.30	1.80	1.20	1.40	1.00	1.30	1.80	1.80	1.60
	12 in.	DN300	1.50	2.10	1.30	1.60	2.00	1.50	2.10	2.10	1.90
	14 in.	DN350	1.75	2.30	1.40	1.80	2.00	1.75	2.30	2.30	2.10
	16 in.	DN400	2.00	2.60	1.50	2.00	2.00	2.00	2.60	2.60	2.40
	18 in.	DN450	2.25	3.00	1.60	2.20	2.00	2.25	3.00	3.00	2.70
	20 in.	DN500	2.50	3.30	1.70	2.40	2.00	2.50	3.30	3.30	3.00
	22 in.	DN550	2.75	3.60	1.80	2.60	3.00	2.75	3.60	3.60	3.15
	24 in.	DN600	3.00	3.90	1.90	2.80	3.00	3.00	3.90	3.90	3.30
	26 in.	DN650	3.20	4.20	2.00	3.00	3.00	3.20	4.20	4.20	3.60
	28 in.	DN700	3.40	4.60	2.10	3.20	3.00	3.40	4.60	4.60	4.20
	30 in.	DN750	3.60	4.90	2.20	3.50	3.00	3.60	4.90	4.90	4.50
	32 in.	DN800	3.80	5.30	2.30	3.70	3.00	3.80	5.30	5.30	4.80
	34 in.	DN850	4.00	5.70	2.40	3.90	3.00	4.00	5.70	5.70	5.10
	36 in.	DN900	4.20	6.10	2.50	4.10	3.00	4.20	6.10	6.10	5.40
	38 in.	DN950	4.40	6.50	2.60	4.30	4.50	4.40	6.50	6.50	5.70
	40 in.	DN1000	4.60	6.90	2.70	4.50	4.50	4.60	6.90	6.90	6.00
	42 in.	DN1050	4.80	7.30	2.80	4.70	4.50	4.80	7.30	7.30	6.30
	44 in.	DN1100	5.00	7.70	2.90	4.90	4.50	5.00	7.70	7.70	6.60
	46 in.	DN1150	5.20	8.10	3.00	5.10	4.50	5.20	8.10	8.10	6.90
	48 in.	DN1200	5.40	8.60	3.10	5.30	4.50	5.40	8.60	8.60	7.20
WALL SCHEDULE	5S, 10S		0.8	0.8	1.0	1.0/n.a./n.a.	1.0		1.0	0.8	
	10, 20, 30, 40, STD		**1.0**	**1.0**	**1.0**	**1.0/n.a./n.a.**	**1.0**		**1.0**	**1.0**	
	60, 80, XS		1.2	1.3	1.0	1.0/n.a./n.a.	1.2		1.0	1.2	
	100, 120, 140, 160, XXS		1.6	2.0	1.5	1.5/n.a./n.a.	1.5		1.0	1.5	
	25.5–50 mm WT		2.2	3.0	3.0	3.0/n.a./n.a.	2.0		1.0	2.0	
FLANGE RATING	**150 lbs**						1.0				
	300 lbs						1.0				
	600 lbs						1.3				
	900 lbs						1.5				
	1500–2500 lbs						2.0				
SPECIAL ITEMS WELD TYPE	WELDED VALVE LARGE BORE			1.5							
	SOCKET WELD			0.6							
	SEAL WELD			0.4							
	ANGLE WELD / DUMMY LEG			1.4							
	BRANCH WELD (90°)			2.5							
	BRANCH WELD (angle not 90°)			3.8							
	REINF. BRANCH WELD (90°)			3.3							
	REINF. BRANCH WELD (not 90°)			4.9							
	WELDING OUTLET (90°)			4.0							
	WELDING OUTLET (not 90°)			6.0							

2.5 Transportation of Piping Materials

The descriptions given below provide a sample of how the Unit Rates for the Transportation of Piping Materials could be handled, depending on locations and subject to prior agreement.

The non-measured items include all equipment and consumables required to execute this work.

1.0001 TRANSPORTATION OF PIPING MATERIALS TO PREFABRICATION SHOP(S)	
Included	The loading at Employer's Site Warehouse or other Warehouse location, Transportation to Contractor's Prefabrication Shop(s) of 'Free Issue' Prefabrication Piping Materials
Excluded	Loading onto Truck (by Employer)
Measurement	Number of Truckloads (loaded to capacity). (*Note: Subject to specific Project Execution Philosophy*)
Unit	Each (ea)
Multipliers	None

1.0002 TRANSPORTATION OF PIPING MATERIALS TO SITE	
Included	The loading at Contractor's Prefabrication Shop(s); Transportation to Site of Piping Materials
Excluded	
Measurement	Number of Truckloads (loaded to capacity). (*Note: Subject to specific Project Execution Philosophy*)
Unit	Each (ea)
Multipliers	None

2.6 Description for Pricing Table 1

(PT-1): Piping Prefabrication and Painting (1.0000 item series)

The descriptions given below describe the Units of Work included in Pricing Table 1. Any supply of Materials 'by Contractor' are to be specified in the Scope Description. The 'non-measured' items include the following:

- Consumable Materials
- Equipment and Tools
- Prefabrication facilities

- Engineering and Coordination/Administration, required for prefabrication
- Quality Assurance/Quality Control.

1.1000 HANDLING

1.1102 HANDLING OF STRAIGHT RUN PIPE

Included	Offloading, Warehousing, Shop Handling, Cleaning and Capping of Straight Run Pipe
	Straight Run Pipe shall be pipe with no In-line fittings welded to it, such as Elbows/Tees/Reducers/Weldneck flanges, etc.
	When only O'lets, any other Branches, pipe supports, reinforcement pads, etc. are welded on a pipe in Prefabrication, it is still considered to be a Straight Run Pipe
	Also, when random length pipe is cut to length and/or beveled in Prefabrication, it is still considered being a Straight Run Pipe
Excluded	Making of Joints, such as Welds
Measurement	Length of Pipe from End to End
Unit	Linear Meter (m¹)
Multipliers	Material Type, Pipe Size, Wall Schedule

1.2000 JOINTS

1.2100 WELDS

Included	Cutting, necessary beveling, pre-heating where required, and welding of Pipe and/or Fittings
	Handling of Pipe Materials: Offloading, Warehousing, Shop Handling, Cleaning and Capping of Pipe Spools, including pipe and all fittings which form part of Pipe Spools
Excluded	Non-Destructive Examination, Hydrostatic Testing, Post-Weld Heat Treatment
Measurement	Number of Welds
Unit	Each (ea)
Multipliers	Material Type, Pipe/Fitting Size, Wall Schedule, Weld Type/Special Item

1.4000 TESTING AND HEAT TREATMENT

1.4101/2 RADIOGRAPHS (X-RAY/GAMMA-RAY)

Included	All work involved with the taking and developing of radiographs, the evaluation, administration and reporting Included is additional handling of Pipe Spools to/from the Radiography Bunker
Excluded	
Measurement	Number of Welds for which radiographs have been approved
Unit	Each (ea)
Multipliers	Pipe Size, Wall Schedule

1.4110 DYE PENETRANT TESTING (DP)

Included	All work involved with the making of a Dye Penetrant Test, the evaluation, administration and reporting Included is additional handling of Pipe Spools
Excluded	
Measurement	Number of welds for which a Dye Penetrant Test have been approved
Unit	Each (ea)
Multipliers	Pipe Size

1.4120 MAGNETIC PARTICLE TESTING (MP)

Included	The taking of Magnetic Particle tests (if required), the evaluation, administration and reporting Included is additional handling of Pipe Spools
Excluded	
Measurement	Number of Welds for which a Magnetic Particle Test have been approved
Unit	Each (ea)
Multipliers	Pipe Size

1.4140 ULTRASONIC TESTING (US)

Included	The taking of Ultrasonic tests (if required), the evaluation, administration and reporting Included is additional handling of Pipe Spools
Excluded	
Measurement	Number of Welds for which an Ultrasonic Test have been approved
Unit	Each (ea)
Multipliers	Pipe Size, Wall Schedule

1.4200 POST-WELD HEAT TREATMENT (PWHT)

Included	All work involved with the Electrical or Furnace Post-Weld Heat Treatment of one Weld. Included is additional Handling of Pipe Spools to/from a Heating Furnace, hardness testing and related documentation
Excluded	
Measurement	Number of Treated Welds
Unit	Each (ea)
Multipliers	Material Type, Pipe Size, Wall Schedule

1.4300 HYDROSTATIC TESTING AND FLUSHING

Included	All Supply of Temporary Materials and all Work involved with the Pre-flushing, Hydrostatic Testing and Flushing of a piping system
	Included are the following activities:
	• opening and closing of Valves • handling and welding/bolt-up of Test Blinds • drain and dry by means of oil free air of ambient temperature after Testing • cutting off of welded Test Blinds and beveling of pipe ends
Excluded	
Measurement	Length of tested lines, centerline to centerline (through all fittings)
Unit	Linear Meter (m^1)
Multipliers	Pipe Size, Flange Rating

1.5000 SUPPORTS

1.5102/3/4/5/6 SUPPLY AND FABRICATION OF LARGE BORE PIPE SUPPORTS

Included	The material supply and fabrication of Carbon Steel pipe supports, guides, hangers, dummy leg base plates; included are all shim-plates required for pipe support base-plates and all bolting materials
	Included are all attachment Welds to pipe, e.g. the welding of shoes to a pipe where required
Excluded	Dummy leg piping materials, for which handling and welds are measured as part of pipe spools
	Field Run Small Bore Supports shall be priced and quantified as Part of the Erection Pricing Table
Measurement	Weight of supports per weight range, net as fabricated, excluding weight of bolts, nuts, washers and shim-plates
Unit	Kilogram (kg)
Multipliers	None

1.6000 PAINTING

Note For materials, number of coats and MDFT (minimum dry film thickness) reference is made to the applicable specification. Included is taping and marking
A Base Unit Rate shall be given for each Paint System

1.6101 PAINTING OF PIPE SPOOLS (INCLUDING FITTINGS)

Included Shop Treatment (blasting/painting) of Prefabricated Pipe Spools and the transportation between the Prefabrication Shop(s) and the Paint Shop

Excluded

Measurement Length of Pipe plus equivalent length of all fittings, flanges, valves

(Note: method of determining equivalent length to be specified by Employer)

Unit Equivalent Linear Meter (m^1)

Multipliers Pipe/Fitting Size

1.6102 PAINTING OF STRAIGHT RUN PIPE

Included Shop Treatment (blasting/painting) of Straight Run Pipe and the transportation between the Prefabrication Shop(s) and the Paint Shop

Excluded

Measurement Length of Pipe from end to end

Unit Linear Meter (m^1)

Multipliers Pipe Size

1.6202/3/4/5/6 PAINTING OF PIPE SUPPORTS

Included Shop Treatment of Pipe Supports and the transportation between the Prefabrication Shop(s) and the Paint Shop

Excluded

Measurement Weight of supports per weight range, net as fabricated, excluding weight of bolts, nuts, washers and shim-plates

Unit Kilogram (kg)

Multipliers None

1.8000 MODIFICATION WORK

1.8010 TAPERING

Included	The Tapering (i.e. by means of inside machining 1:4) of any type of pipe or fitting from existing Wall Schedule to one Schedule down. Adaptations of 2 mm or less shall not be measured. Included is additional handling of pipe or fittings
Excluded	
Measurement	Number of tapered ends
Unit	Each (ea)
Multipliers	Material Type, Pipe/Fitting Size

1.8020 DRILLING/THREADING

Included	The drilling of a hole and threading in blinds, etc. Included is the handling of the blind
Excluded	
Measurement	Number of threaded holes made
Unit	Each (ea)
Multipliers	None

1.8040 THREADING

Included	The cutting of pipe and the threading of the pipe end
Excluded	
Measurement	Number of threaded ends made
Unit	Each (ea)
Multipliers	Pipe Size (refer to cut/bevel table)

1.8101 DOUBLE HANDLING OF PIPE SPOOLS (FOR MODIFICATION WORK)

Included	The Handling of an already fabricated Pipe Spool necessary for modifications thereof due to drawing revisions
Excluded	Making of Joints, such as Welds
Measurement	Length of Pipe Spools from centerline to centerline through all fittings
Unit	Linear Meter (m¹)
Multipliers	Material Type, Pipe/Fitting Size, Wall Schedule

1.8301 CUT (FOR MODIFICATION WORK)

Included	The cutting of already fabricated Pipe Spools necessary for modifications thereof due to drawing revisions
Excluded	
Measurement	Number of cuts made in addition to cuts already included in additional Welds necessary for modifications to already fabricated Pipe Spools (included in the Unit Rate for a Weld is one cut)
Unit	Each (ea)
Multipliers	Material Type, Pipe Size, Wall Schedule

1.8302 BEVEL (FOR MODIFICATION WORK)

Included	The beveling after cutting of already fabricated Pipe Spools necessary for modifications thereof due to drawing revisions
Excluded	
Measurement	Number of beveled ends made in addition to bevels already included in additional Welds necessary for modifications to already fabricated Pipe Spools (included in the Unit Rate for a Weld are two bevels)
Unit	Each (ea)
Multipliers	Material Type, Pipe Size, Wall Schedule

1.8400 WELDS (FOR MODIFICATION WORK)

Included	Cutting, beveling (both ends) and welding of Pipe and/or Fittings
Excluded	Non-destructive Testing, Hydrostatic Testing or Heat Treatment
Measurement	Number of Welds
Unit	Each (ea)
Multipliers	Material Type, Pipe/Fitting Size, Wall Schedule

2.7 Description for Piping Erection Multipliers (Multiplier Table MT-2)

2.7.1 Multipliers for Material Type

The Base Unit Rates for Handling, Welds, Screwed Joints, Socket Weld Joints, Bends, Heat Treatment, Tapering, Cuts and Bevels are based on:

Carbon Steel

(for which the Multiplier is set as *1.0*)

The Multipliers for all other Material types, as listed in Multiplier Table MT-2, shall be applied to the above mentioned Base Unit Rates to cover variations to the extent of the Work involved with the applicable Material types.

2.7.2 Multipliers for Pipe/Fitting Size

The Base Unit Rates for Handling, Welds, Flanged Joints, Screwed Joints, Bends, NDE, PWHT, Tapering, Cuts and Bevels, and for Painting are based on:

For Large Bore Piping: Outside Diameter 6 in. (or DN150)
For Small Bore Piping: Outside Diameter 3/4 inch (or DN20)

(for which the Multipliers are set as *1.0*)

The Multipliers for Pipe/Fitting Size for all other outside diameters, as listed in Multiplier Table MT-2, shall be applied to the above mentioned Base Unit Rates to cover variations to the extent of the Work involved with the applicable outside diameters.

2.7.3 Multipliers for Pipe Wall Schedule

The Base Unit Rates for Handling, Welds, Bends, Screwed Joints, Socket Weld Joints, NDE, PWHT, Tapering, Cuts and Bevels are based on:

For Large Bore Piping: Wall Schedule STANDARD (STD)
For Small Bore Piping: Schedule 80

(for which the Multiplier is set as *1.0*)

The Multipliers for Pipe Wall Schedule for all other Wall Schedules, as listed in Multiplier Table MT-2, shall be applied to the above mentioned Base Unit Rates to cover variations to the extent of the Work involved with the applicable wall schedules.

For any wall-thickness between 1 in. (25.5 mm) and 2 in. (50 mm), the multiplier for '25.5–50 mm wall thickness' shall prevail.

2.7.4 Multipliers for Pipe Flange Rating

The Base Unit Rate for Handling of In-line Items and Flanged Joints is based on:

Flange Rating 150 lbs

(for which the Multiplier is set as *1.0*)

The Multipliers for Flange Rating for all other Flange Ratings, as listed in Multiplier Table MT-2, shall be applied to the above mentioned Base Unit Rate to cover variations to the extent of the Work involved with the applicable flange ratings.

2.7.5 Multipliers for Weld Types/Special Items

The Base Unit Rates for Weld Types are based on:

Butt Weld/Fillet Weld

(for which the Multiplier is set as *1.0*)

The Multipliers for all other Weld Types, as listed in Multiplier Table MT-2, shall be applied to the above mentioned Base Unit Rate to cover variations to the extent of the Work involved with the applicable Weld types. Note that for Welds on slip-on flanges, Multiplier *1.0* shall also be used.

2.7.5.1 Multiplier for Large Bore Welded Valves

The Multiplier for Large Bore Welded Valves applies to the Base Unit Rate for Handling In-line Items and covers the extra work involved with dismantling, handling and mounting of valve internals and/or special handling of spools containing welded-in valves.

2.7.5.2 Multiplier for Socket Welds

The Multiplier for welding Socket Welds applies to the Base Unit Rate for Welds and covers the reduced work involved with welding a Socket Weld as against a Butt Weld.

2.7.5.3 Multiplier for Seal Welds

The Multiplier for welding Seal Welds applies to the Base Unit Rate for Welds and covers the reduced work involved with welding a Seal Weld as against a Butt Weld.

2.7.5.4 Multiplier for Angle Welds/Dummy Legs

The Multiplier for welding Angle Welds and/or Dummy Legs applies to the Base Unit Rate for Welds and covers the extra work involved with welding an Angle Weld for a mitre-bend, or for welding a Dummy Leg to a pipe.

2.7.5.5 Multiplier for Branch Welds (90°)

The Multiplier for Branch Welds (including nipples, half-couplings and bosses) applies to the Base Unit Rate for Welds of the branch size and covers the extra work involved with drilling/cutting a hole in, and welding a branch to, the main pipe under a straight angle of 90°.

2.7.5.6 Multiplier for Branch Welds (angle not 90°)

The Multiplier for Branch Welds applies to the Base Unit Rate for Welds of the branch size and covers the extra work involved with drilling/cutting a hole in, and welding a branch to, the main pipe under an angle that is not 90°.

2.7.5.7 Multiplier for Reinforced Branch Welds (90°)

The Multiplier for Reinforced Branch Welds applies to the Base Unit Rate for Welds of the branch size. It covers the extra work involved with drilling/cutting a hole in, and welding of a branch to, the main pipe under a straight angle of 90°, as well as fabrication from pipe, handling and welding a Branch Reinforcement Pad.

2.7.5.8 Multiplier for Reinforced Branch Welds (angle not 90°)

The Multiplier for Reinforced Branch Welds applies to the Base Unit Rate for Welds of the branch size. It covers the extra work involved with drilling/cutting a hole in, and welding a branch to, the main pipe under an angle that is not 90°, as well as fabrication from pipe, handling and welding a Branch Reinforcement Pad.

2.7.5.9 Multiplier for Welding Outlets (90°)

The Multiplier for Welding Outlets (Weldolets, Flangolets, Sockolets, Nippolets, Threadolets) applies to the Base Unit Rate for Welds of the Outlet size. It covers the extra work involved with drilling/cutting a hole in, and welding an Outlet to, the main pipe under a straight angle that is 90°.

2.7.5.10 Multiplier for Welding Outlets (angle not 90°)

The Multiplier for Welding Outlets (Latrolets, Elbolets) applies to the Base Unit Rate for Welds of the Outlet size and covers the extra work involved with welding an Outlet to the main pipe under an angle that is not 90°.

2.7.5.11 Multiplier for Control/Safety Valves

The Multiplier for Control/Safety Valves applies to the Base Unit Rate for Handling Flanged In-line Items and covers the extra work involved with dismantling, checking, handling and mounting of Valve internals.

ECI Multiplier Table MT-2: PIPING ERECTION UNIT RATE MULTIPLIER TABLE

MATERIAL TYPE

Material Type	HANDLING	WELD JOINT	HANDLING IN-LINE ITEMS	FLANGED JOINT	SCREWED JOINT	BEND	NDE	PWHT	HYDRO TEST	TAPERING	CUTS/BEVELS	PAINTING
CARBON STEEL	1.0	1.0	1.0		1.0	1.0	1.0	1.0	1.0	1.0		
KILLED CARBON STEEL	1.0	1.0	1.0		1.0	1.0	1.0	1.0	1.0	1.0	1.0	
STAINLESS STEEL	1.1	1.6	1.0		1.2	1.3	1.0		1.0	1.2	1.2	
ALLOY STEEL	1.0	2.0	1.0		1.2	1.3	1.0	1.2	1.0	1.2	1.2	

PIPE/FITTING SIZE

Size	DN	HANDLING	WELD JOINT	HANDLING IN-LINE ITEMS	FLANGED JOINT	SCREWED JOINT	BEND	NDE	PWHT	HYDRO TEST	TAPERING	CUTS/BEVELS	PAINTING
1/2 in	DN15	0.80	0.80	0.80	0.80	0.80	0.80	1.00	1.00	0.80	0.80	0.80	0.30
3/4 in	DN20	1.00	1.00	1.00	1.00	1.00	1.00	1.00	1.00	1.00	1.00	1.00	0.30
1 in	DN25	1.10	1.10	1.10	1.10	1.10	1.10	1.00	1.00	1.10	1.10	1.10	0.30
11/2 in	DN40	1.40	1.40	1.40	1.40	1.40	1.40	1.00	1.00	1.40	1.40	1.40	0.30
2 in	DN50	0.70	0.60	0.55	0.65			0.80	1.00	0.70	0.60	0.60	0.40
3 in	DN80	0.80	0.70	0.70	0.70			0.85	1.00	0.80	0.70	0.70	0.55
4 in	DN100	0.90	0.80	0.80	0.80			0.90	1.00	0.90	0.80	0.80	0.70
6 in	DN150	1.00	1.00	1.00	1.00			1.00	1.00	1.00	1.00	1.00	1.00
8 in	DN200	1.10	1.40	1.25	1.25			1.15	1.00	1.10	1.40	1.40	1.30
10 in	DN250	1.30	1.80	1.50	1.50			1.30	1.00	1.30	1.80	1.80	1.60
12 in	DN300	1.50	2.10	1.75	1.75			1.45	2.00	1.50	2.10	2.10	1.90
14 in	DN350	1.75	2.30	2.00	2.00			1.60	2.00	1.75	2.30	2.30	2.10
16 in	DN400	2.00	2.60	2.25	2.25			1.75	2.00	2.00	2.60	2.60	2.40
18 in	DN450	2.25	3.00	2.50	2.50			1.90	2.00	2.25	3.00	3.00	2.70
20 in	DN500	2.50	3.30	2.75	2.75			2.05	2.00	2.50	3.30	3.30	3.00
22 in	DN550	2.75	3.60	3.00	3.00			2.20	3.00	2.75	3.60	3.60	3.15
24 in	DN600	3.00	3.90	3.25	3.25			2.35	3.00	3.00	3.90	3.90	3.30
26 in	DN650	3.20	4.20	3.50	3.50			2.50	3.00	3.20	4.20	4.20	3.60
28 in	DN700	3.40	4.60	3.75	3.75			2.65	3.00	3.40	4.60	4.60	4.20
30 in	DN750	3.60	4.90	4.00	4.00			2.80	3.00	3.60	4.90	4.90	4.50
32 in	DN800	3.80	5.30	4.30	4.25			2.95	3.00	3.80	5.30	5.30	4.80
34 in	DN850	4.00	5.70	4.60	4.50			3.10	3.00	4.00	5.70	5.70	5.10
36 in	DN900	4.20	6.10	5.00	4.75			3.25	3.00	4.20	6.10	6.10	5.40
38 in	DN950	4.40	6.50	5.40	5.00			3.40	4.50	4.40	6.50	6.50	5.70
40 in	DN1000	4.60	6.90	5.80	5.25			3.55	4.50	4.60	6.90	6.90	6.00
42 in	DN1050	4.80	7.30	6.20	5.50			3.70	4.50	4.80	7.30	7.30	6.30
44 in	DN1100	5.00	7.70	6.60	5.75			3.85	4.50	5.00	7.70	7.70	6.60
46 in	DN1150	5.20	8.10	7.00	6.00			4.00	4.50	5.20	8.10	8.10	6.90
48 in	DN1200	5.40	8.60	7.40	6.25			4.15	4.50	5.40	8.60	8.60	7.20

(1/2 in – 11/2 in = SMALL BORE; 2 in and above = LARGE BORE)

WALL SCHEDULE

Schedule	HANDLING	WELD JOINT	HANDLING IN-LINE ITEMS	FLANGED JOINT	SCREWED JOINT	BEND	NDE	PWHT	HYDRO TEST	TAPERING	CUTS/BEVELS	PAINTING
SMALL BORE												
5S, 10S	0.8	0.8			1.0	1.0	1.0	1.0		1.0	1.0	
40, STD	0.9	0.9			1.0	1.0	1.0	1.0		1.0	1.0	
80, XS	1.0	1.0			1.0	1.0	1.0	1.0		1.0	1.0	
160, XXS	1.2	1.2			1.2	1.3	1.0	1.0		1.0	1.0	
LARGE BORE												
5S, 10S	0.8	0.8					1.0	1.0		1.0	0.8	
10, 20, 30, 40, STD	1.0	1.0					1.0	1.0		1.0	1.0	
60, 80, XS	1.2	1.3					1.0	1.2		1.0	1.2	
100, 120, 140, 160, XXS	1.6	2.0					1.5	1.5		1.0	1.5	
25.5–50 mm WT	2.2	3.0					3.0	2.0		1.0	2.0	

FLANGE RATING

Rating	HANDLING IN-LINE ITEMS	FLANGED JOINT	HYDRO TEST
150 lbs	1.0	1.0	1.0
300 lbs	1.15	1.15	1.0
600 lbs	1.3	1.3	1.3
900 lbs	1.5	1.5	1.5
1500–2500 lbs	2.0	2.0	2.0

SPECIAL ITEMS / WELD TYPE

Item	HANDLING IN-LINE ITEMS	WELD JOINT
WELDED VALVE LARGE BORE	2.0	
SOCKET WELD		0.6
SEAL WELD		0.4
ANGLE WELD/DUMMY LEG		1.4
BRANCH WELDS (90°)		2.5
BRANCH WELD (angle not 90°)		3.8
REINF. BRANCH WELDS (90°)		3.3
REINF. BRANCH WELDS (not 90°)		4.9
WELDING OUTLET (90°)		4.0
WELDING OUTLET (non 90°)		6.0
CONTROL VALVE	2.0	
SAFETY VALVE	1.5	

Note 1: For NDE, Wall Schedule Factors shall not be applied for DP and MPT.

2.8 Description for Pricing Table 2

(PT-2): Piping Erection (2.0000 item series)

The descriptions given below describe the Units of Work included in Pricing Table 2. Any supply of Materials 'by Contractor' are to be specified in the Scope Description. The 'non-measured' items include the following:

- the supply, installation and removal of temporary erection bolts, nuts, gaskets, flanges, blinds, spades, plates, etc.
- the supply, installation/use (including maintenance)/calibration and removal of testing materials and/or equipment
- the supply, installation and removal of temporary supporting measures.

2.1000 HANDLING
2.1100 HANDLING OF PIPE
2.1101 HANDLING OF PREFABRICATED PIPE SPOOLS

Included	Offloading, Temporary Storage, On-site Transportation, Handling and Erection, Level and Alignment of prefabricated Pipe Spools
	Only when In-line fittings, such as Elbows, Tees, Reducers, Weld-neck Flanges etc., are included, shall piping be considered to be a Pipe Spool
Excluded	Making of Joints, such as Welds and Bolt-up of Flanges, Hydrostatic Testing
Measurement	Length of Pipe Spools from centerline to centerline through all fittings, branches, dummy legs, etc.
Unit	Linear Meter (m¹)
Multipliers	Material type, Pipe/Fitting Size, Wall Schedule

Note: This item covers Large Bore prefabricated pipe spools as well as Small Bore, either shop or prefabricated

2.1102 HANDLING OF STRAIGHT RUN PIPE

Included	Offloading, Temporary Storage, On-site Transportation, Handling and Erection, Level and Alignment of Straight Run Pipe
	Straight Run Pipe shall be a pipe with no In-line fittings welded to such a pipe, such as Elbows, Tees, Reducers, Weld-neck Flanges, etc.
	When only O'lets, any other Branches, pipe supports, reinforcement pads, etc. are welded on a pipe in Prefabrication, it is still considered to be a Straight Run Pipe
	Also, when random length pipe is cut to length and/or beveled in Prefabrication, it is still considered to be a Straight Run Pipe
Excluded	Making of Joints, such as Welds and Bolt-up of Flanges, Hydrostatic Testing
Measurement	Length of Pipe from end to end
Unit	Linear Meter (m^1)
Multipliers	Material Type, Pipe Size, Wall Schedule

2.1200 HANDLING OF NON-WELDED IN-LINE ITEMS

Included	Loading at Warehouse, On-site Transportation, Temporary Storage, Handling and Erection, Level and Alignment of non-welded In-line items, such as, but not limited to, flanged valves, strainers, silencers, bellows, gauges, thermowells, etc.
Excluded	Bolt-up of Flanges, making of screwed connections, Hydrostatic Testing
Measurement	Number of non-welded In-line items
Unit	Each (ea)
Multipliers	Pipe Size, Rating, Special Items

2.1300 HANDLING (for certification) OF SAFETY/RELIEF VALVES

Included	Dismantling, Handling, Transportation (Testing, Verification, Certification by an approved and authorized party) and Reinstatement of Safety/Relief Valves
Excluded	Bolt-up of Flanges
Measurement	Number of Safety/Relief Valves certified
Unit	Each (ea)
Multipliers	None

2.2000 JOINTS

2.2100 FIELD WELDS

Included	Cutting, necessary beveling, pre-heating where required, and welding of pipe and/or fittings
	Included is the protection of surrounding surfaces to avoid damages caused by grinding and welding
Excluded	Non-destructive Examination, Hydrostatic Testing or PWHT
Measurement	Number of welds
Unit	Each (ea)
Multipliers	Material Type, Pipe Size, Wall Schedule, Special Item/Weld Type

2.2200 FLANGED JOINTS

Included	The Bolt-up of a Flanged Joint, including the handling of bolts, nuts, washers, gaskets, rings, spades, spectacle blinds and blind flanges
	Included are Bolt Tensioning (torque-wrenching) and the administration thereof, if required
Excluded	Handling of Pipe and Fittings, Hydrostatic Testing
Measurement	Number of Flanged Joints. When rings or spades are inserted between two Flanges, one Flanged Joint shall be measured
Unit	Each (ea)
Multipliers	Pipe Size, Flange Rating

2.2250 HYDRAULIC BOLT TENSIONING OF FLANGED JOINTS

Included	The Hydraulic Tensioning of a Flanged Joint, including the administration thereof, all in accordance with the applicable specifications
Excluded	Flanged Joint
Measurement	Number of Hydraulically Tensioned Flanged Joints
Unit	Each (ea)
Multipliers	Pipe Size, Flange Rating (refer to Flanged Joint Table)

2.2300 SCREWED JOINTS

Included	The cutting and threading of pipe and the making of a Screwed Joint
Excluded	Handling of Pipe and Fittings, Hydrostatic Testing
Measurement	Number of Screwed Joints
Unit	Each (ea)
Multipliers	Material Type, Pipe Size, Wall Schedule

2.3000 BENDS

Included	The bending of a pipe
Excluded	Handling of Pipe
Measurement	Number of bends made
Unit	Each (ea)
Multipliers	Material Type, Pipe Size, Wall Schedule

2.4000 TESTING AND TREATMENT
2.4101/2 RADIOGRAPHS (X-RAY/GAMMA-RAY)

Included	All work involved with the taking and developing of radiographs, the evaluation, administration and reporting
Excluded	
Measurement	Number of Welds for which radiographs have been approved
Unit	Each (ea)
Multipliers	Pipe Size, Wall Schedule

2.4110 DYE PENETRANT TESTING (DP)

Included	All work involved with the making of a Dye Penetrant Test, the evaluation, administration and reporting
Excluded	
Measurement	Number of Welds for which a Dye Penetrant Test have been approved
Unit	Each (ea)
Multipliers	Pipe Size

2.4120 MAGNETIC PARTICLE TESTING (MP)

Included	The taking of Magnetic Particle tests, the evaluation, administration and reporting
Excluded	
Measurement	Number of Welds for which a Magnetic Particle Test have been approved
Unit	Each (ea)
Multipliers	Pipe Size

2.4140 ULTRASONIC TESTING (US)

Included   The taking of Ultrasonic tests, the evaluation, administration and reporting

Excluded

Measurement   Number of Welds for which an Ultrasonic Test have been approved

Unit   Each (ea)

Multipliers   Pipe Size, Wall Schedule

2.4200 POST-WELD HEAT TREATMENT (PWHT)

Included   All work involved with the Post-Weld Heat Treatment of one Weld

Excluded

Measurement   Number of Treated Welds

Unit   Each (ea)

Multipliers   Pipe Size, Wall Schedule

2.4300 HYDROSTATIC TESTING, FLUSHING AND REINSTATEMENT

Included   All Supply of Temporary Materials and all Work involved with the Cleaning, Pre-flushing, Hydrostatic Testing, Flushing, Draining, Drying and Reinstatement of a piping system.

Included are the following activities:

- spading, de-spading
- breaking of Flanged Joints
- opening and closing of Valves
- blocking and unblocking of spring type supports/hangers
- handling and Welding/Bolt-up of Test Blinds
- drain and dry by means of oil free air of ambient temperature after Testing
- cutting off of welded Test Blinds and beveling of pipe ends
- bolt-up and torque-wrenching (if required) of Flanged Joints

Excluded   Supply of Test and Flush Water and disposal thereof (if off site)

Measurement   Length of tested lines centerline to centerline (through all fittings)

Unit   Linear Meter (m^1)

Multipliers   Pipe Size, Flange Rating

2.5000 SUPPORTS

2.5101 PREFABRICATION AND INSTALLATION OF SMALL BORE FIELD RUN PIPE SUPPORTS (<2 IN.)

Included	Included is the design, if required
	The material supply and fabrication of Carbon Steel pipe supports, guides, hangers, dummy leg base-plates. Included are all shim-plates required for pipe support base-plates and all bolting materials
	Included are all attachment welds to a pipe, e.g. the welding of shoes to a pipe where required
	Offloading, Temporary Storage, On-site Transportation, Handling and Erection of pipe supports, drilling, guides, hangers, dummy leg base-plates. Included are all shim-plates required for pipe support base-plates and all bolting materials
Excluded	Dummy leg piping materials, for which Handling and Welds are measured as part of Pipe Spools. Grouting of supports
Measurement	Weight of supports per weight range, net as fabricated, excluding weight of bolts, nuts, washers and shim-plates
Unit	Kilogram (kg)
Multipliers	None

2.5102/3/4/5/6 INSTALLATION OF LARGE BORE PIPE SUPPORTS

Included	Offloading, Temporary Storage, On-Site Transportation, Handling and Erection of pipe supports, drilling, guides, hangers, dummy leg base-plates. Included are all shim-plates required for pipe support base-plates and all bolting materials
Excluded	Dummy leg piping materials, for which Handling and Welds are measured as part of Pipe Spools. Grouting of supports
Measurement	Weight of supports per weight range, net as fabricated, excluding weight of bolts, nuts, washers and shim-plates
Unit	Kilogram (kg)
Multipliers	None

2.5301/2/3 SPRING HANGERS

Included	Offloading, Site Transportation, Warehousing, Handling and Erection of Spring Hanger assemblies and spring type supports
Excluded	
Measurement	Number of Spring Hangers per weight range
Unit	Each (ea)
Multipliers	None

2.6000 FIELD PAINTING

Note For materials, number of coats and MDFT (minimum dry film thickness), reference is made to the applicable specification. Included is taping and marking

2.6100 CLEANING AND PAINTING OF WELDS

Included All work involved with the surface preparation, priming and finishing, including the area adjacent to the Weld.

Included are the required protection of all surrounding surfaces to avoid contamination and all the required weather protection to ensure correct humidity and temperature environment for painting

Excluded

Measurement Number of painted Welds

Unit Each (ea)

Multipliers Pipe/Fitting Size

2.6101 FINAL PAINTING OF INSTALLED PIPE

Included Field treatment of installed pipe, including all fittings, valves, etc.

Excluded

Measurement Length of Pipe plus equivalent length of all fittings, flanges, valves

(Note: method of determining equivalent length to be specified by Employer)

Unit Equivalent Linear Meter (m^1)

Multipliers Pipe/Fitting Size

2.6201/2/3/4/5/6 FIELD PAINTING OF PIPE SUPPORTS

Included Field treatment of Pipe Supports, including all handling

Excluded

Measurement Weight of supports per weight range, net as fabricated, excluding weight of bolts, nuts, washers and shim-plates

Unit Kilogram (kg)

Multipliers None

2.8000 MODIFICATION WORK

2.8010 TAPERING

Included	The Tapering (i.e. by means of inside machining 1:4) of any type of pipe or fitting from existing Wall Schedule to one Schedule down. Adaptations of 2 mm or less shall not be measured.
	Included is the additional handling of pipe or fittings.
Excluded	
Measurement	Number of tapered ends
Unit	Each (ea)
Multipliers	Material Type, Pipe/Fitting Size

2.8020 DRILLING/THREADING

Included	The Drilling of a hole and Threading in blinds, etc. Included is the handling of the blind
Excluded	
Measurement	Number of threaded holes made
Unit	Each (ea)
Multipliers	None

2.8040 THREADING

Included	The Cutting of pipe and Threading of the pipe end
Excluded	
Measurement	Number of threaded ends made
Unit	Each (ea)
Multipliers	Pipe Size (refer to cut/bevel table)

2.8101 DOUBLE HANDLING OF PIPE SPOOLS (FOR MODIFICATION WORK)

Included	The Dismantling, Handling and Reinstatement of already erected Pipe Spools necessary for modifications thereof due to drawing and/or field revisions.
	Included is the removal of any redundant piping materials
Excluded	The making of Joints, such as Welds and Bolt-up of Flanges
Measurement	Length of Pipe Spools from centerline to centerline through all fittings
Unit	Linear Meter (m^1)
Multipliers	Pipe/Fitting Size, Wall Schedule

2.8102 DOUBLE HANDLING OF STRAIGHT RUN PIPE (FOR MODIFICATION WORK)

Included	The Dismantling, Handling and Reinstatement of already erected Straight Run Pipe necessary for modifications thereof due to drawing and/or field revisions.
	Included is the removal of any redundant piping materials
Excluded	The making of Joints, such as Welds and Bolt-up of Flanges
Measurement	Length of Pipe from end to end
Unit	Linear Meter (m^1)
Multipliers	Pipe Size, Wall Schedule

2.8103 DOUBLE HANDLING OF IN-LINE ITEMS (FOR MODIFICATION WORK)

Included	The Handling of already erected In-line Items necessary for modifications thereof due to drawing and/or field revisions, e.g. turning of already installed Flanged Valves
Excluded	The making of Joints, such as Bolt-up of Flanges
Measurement	Number of items handled
Unit	Each (ea)
Multipliers	Pipe Size, Flange Rating

2.8301 CUT (FOR MODIFICATION WORK)

Included	Cutting of already fabricated Pipe Spools necessary for modifications thereof due to drawing and/or field revisions
Excluded	
Measurement	Number of cuts made in addition to cuts already included in additional Welds necessary for modifications to already fabricated Pipe Spools (included in the Unit Rate for a Weld is one cut)
Unit	Each (ea)
Multipliers	Material Type, Pipe Size, Wall Schedule

2.8302 BEVEL (FOR MODIFICATION WORK)

Included	The beveling after cutting of already fabricated Pipe Spools necessary for modifications thereof due to drawing and/or field revisions
Excluded	
Measurement	Number of beveled ends made in addition to bevels already included in additional Welds necessary for modifications to already fabricated Pipe Spools (included in the Unit Rate for a Weld are two bevels)
Unit	Each (ea)
Multipliers	Material Type, Pipe Size, Wall Schedule

2.8400 FIELD WELDS (FOR MODIFICATION WORK)

Included	Cutting, beveling (both ends) and welding of pipe and/or fittings
Excluded	Non-Destructive Examination, Hydrostatic Testing, PWHT
Measurement	Number of Welds
Unit	Each (ea)
Multipliers	Material Type, Pipe Size, Wall Schedule

2.8500 FLANGED JOINTS (FOR MODIFICATION WORK)

Included	The Bolt-up of a Flange Joint, including the handling of bolts, nuts, washers, gaskets, rings, spades and blinds
	Included is Bolt Tensioning (torque-wrenching) and the administration thereof, if required
Excluded	
Measurement	Number of Flanged Joints. When rings or spades are inserted between two Flanges, one Flanged Joint shall be measured
Unit	Each (ea)
Multipliers	Pipe Size, Flange Rating

2.8550 BREAKING FLANGED JOINTS (FOR MODIFICATION WORK)

Included	The breaking (loosening) of a Flanged Joint, including the handling of bolts, nuts, washers, gaskets, rings and spades, necessary for modifications to piping due to drawing and/or field revisions
Excluded	
Measurement	Number of Flanged Joints dismantled. When rings or spades are inserted between two Flanges, one Flanged Joint shall be measured
Unit	Each (ea)
Multipliers	Pipe Size, Flange Rating

3. Pricing Tables: Prefabrication (PT-1)

Pricing Table 1: **Piping Prefabrication and Painting**
Summary of Base Unit Rates

Unit No.	Description		
	Piping Prefabrication and Painting	Currency:	

1.0000	Transportation	Unit	Base Unit Rate
1.0001	Transportation of Piping Materials to Prefabrication Shop(s)	ea	
1.0002	Transportation of Piping Materials to Site	ea	

1.1000	Handling	Unit	Base Unit Rate
1.1102	Handling of Straight Run Pipe	m^1	

1.2000	Joints	Unit	Base Unit Rate
1.2100	Welds	ea	

1.4000	Testing and Heat Treatment	Unit	Base Unit Rate
1.4101	Radiographs (X-Ray)	ea	
1.4102	Radiographs (Gamma-Ray)	ea	
1.4110	Dye Penetrant Testing (DP)	ea	
1.4120	Magnetic Particle Testing (MP)	ea	
1.4140	Ultrasonic Testing (US)	ea	
1.4200	Post-Weld Heat Treatment (PWHT)	ea	
1.4300	Hydrostatic Testing and Flushing	ea	

1.5000	Supply and Fabrication of Large Bore Pipe Supports	Unit	Base Unit Rate
1.5102	Large Bore Pipe Supports (0–5 kg)	kg	
1.5103	Large Bore Pipe Supports (5–10 kg)	kg	
1.5104	Large Bore Pipe Supports (10–20 kg)	kg	
1.5105	Large Bore Pipe Supports (20–50 kg)	kg	
1.5106	Large Bore Pipe Supports (>50 kg)	kg	

1.6000	Painting	Unit	Base Unit Rate
1.6101	Painting of Pipe Spools (Including Fittings)	m^1	
1.6102	Painting of Straight Run Pipe	m^1	
1.6202	Painting of Pipe Supports (0–5 kg)	kg	
1.6203	Painting of Pipe Supports (5–10 kg)	kg	
1.6204	Painting of Pipe Supports (10–20 kg)	kg	
1.6205	Painting of Pipe Supports (20–50 kg)	kg	
1.6206	Painting of Pipe Supports (>50 kg)	kg	

1.8000	Modification Work	Unit	Base Unit Rate
1.8010	Tapering	ea	
1.8020	Drilling/Threading	ea	
1.8040	Threading	ea	
1.8101	Double Handling of Pipe Spools (for Modification Work)	m^1	
1.8301	Cut (for Modification Work)	ea	
1.8302	Bevel (for Modification Work)	ea	
1.8400	Welds (for Modification Work)	ea	

Pricing Table 1: **Transportation of Piping Materials**

Unit No.	Description			
	Transportation to Prefabrication Shop(s)			
		Quantity	Unit Rate	Total Amount
1.0001	Transportation of Piping Materials to Prefabrication Shop(s)			
	Provisional Total Transportation of Piping Materials to Prefabrication Shop(s)			

Unit No.	Description			
	Transportation to Site			
		Quantity	Unit Rate	Total Amount
1.0002	Transportation of Piping Materials to Site			
	Provisional Total Transportation of Piping Materials to Site			

Pricing Table 1: **Handling of Straight Run Pipe**

| Material Type | Size | Schedule | Pipe Length (m) | Base Unit Rate | Multipliers | | | Total Amounts |
					Material Type	Size	Schedule	
1.1102		Totals						

Pricing Table 1: **Welds**

						Multipliers				
Material Type	Size	Schedule	Special Items	No. of Welds	Base Unit Rate	Material Type	Size	Schedule	Special Items	Total Amounts
1.2100			Totals							

Pricing Table 1: **Radiographs (X-Ray)**

| Size | Schedule | No. of X-Rays | Base Unit Rate | Multipliers | | Total Amounts |
				Size	Schedule	
1.4101	Totals					

Pricing Table 1: **Radiographs (Gamma-Ray)**

| Size | Schedule | No. of Gamma-Rays | Base Unit Rate | Multipliers | | Total Amounts |
				Size	Schedule	
1.4102	**Totals**					

Pricing Table 1: **Dye Penetrant Testing (DP)**

| Size | No. of DP Tests | Base Unit Rate | Multipliers | |
			Size	Total Amounts
1.4110	**Totals**			

Pricing Table 1: **Magnetic Particle Testing (MP)**

Size	No. of MP Tests	Base Unit Rate	Multipliers	
			Size	Total Amounts
1.4120 Totals				

Pricing Table 1: **Ultrasonic Testing (US)**

| Size | Schedule | No. of US Tests | Base Unit Rate | Multipliers | | Total Amounts |
				Size	Wall Schedule	
1.4140	Totals					

Pricing Table 1: **Post-Weld Heat Treatment (PWHT)**

Material Type	Size	Schedule	No. of Welds for Treatment	Base Unit Rate	Multipliers Material Type	Size	Schedule	Total Amounts
1.4200		Totals						

Pricing Table 1: **Hydrostatic Testing and Flushing**

| Size | Flange Rating | Line Length (through all fittings) | Base Unit Rate | Multiplier | | Total Amounts |
				Size	Flange Rating	
1.4300	Totals					

Pricing Table 1: **Supply and Fabrication of Large Bore Pipe Supports**

1.5000	SUPPORTS	Weight (kg)	Base Unit Rate	Total Amounts
1.5102	Large Bore Pipe Supports (0–5 kg)			
1.5103	Large Bore Pipe Supports (5–10 kg)			
1.5104	Large Bore Pipe Supports (10–20 kg)			
1.5105	Large Bore Pipe Supports (20–50 kg)			
1.5106	Large Bore Pipe Supports (>50 kg)			
1.5000	Subtotal SUPPORTS			

Pricing Table 1: **Painting of Pipe Spools (including Fittings)**

Paint System	Size	Spool Length (m)	Base Unit Rate	Multiplier Size	Total Amounts
1.6101	**Totals**				

Pricing Table 1: **Painting of Straight Run Pipe**

Paint System	Size	Pipe Length (m)	Base Unit Rate	Multiplier		Total Amounts
				Size		
1.6102	**Totals**					

Pricing Table 1: **Painting of Pipe Supports**

1.6200	PAINTING OF PIPE SUPPORTS	Weight (kg)	Base Unit Rate	Total Amounts
1.6202	Painting of Pipe Supports (0–5 kg)			
1.6203	Painting of Pipe Supports (5–10 kg)			
1.6204	Painting of Pipe Supports (10–20 kg)			
1.6205	Painting of Pipe Supports (20–50 kg)			
1.6206	Painting of Pipe Supports (>50 kg)			
1.6200	Subtotal PAINTING			

Pricing Table 1: **Tapering**

| Material Type | Size | No. of Taperings | Base Unit Rate | Multipliers | | Total Amounts |
				Material Type	Size	
1.8010	**Totals**					

Pricing Table 1: **Drilling/Threading**

		Quantity	Base Unit Rate	Total Amounts
1.8020	**Drilling/Threading**			
	Totals			

Pricing Table 1: **Threading**

Size	No. of Threadings	Base Unit Rate	Multipliers Size	Total Amounts
1.8040 **Totals**				

Pricing Table 1: **Double Handling of Pipe Spools** (for modification work)

Material Type	Size	Schedule	Pipe Length (m)	Base Unit Rate	Multipliers			Total Amounts
					Material Type	Size	Schedule	
1.8101		Totals						

Pricing Table 1: **Cut** (for modification work)

Material Type	Size	Schedule	Number of Cuts	Base Unit Rate	Multipliers Material Type	Size	Schedule	Total Amounts
1.8301		Totals						

Pricing Table 1: **Bevel** (for modification work)

Material Type	Size	Schedule	Number of Bevels	Base Unit Rate	Material Type	Size	Schedule	Total Amounts
						Multipliers		
1.8302		Totals						

Pricing Table 1: **Welds** (for modification work)

					Multipliers			
Material Type	Size	Schedule	Number of Welds	Base Unit Rate	Material Type	Size	Schedule	Total Amounts
1.8400		Totals						

4. Pricing Tables: Erection (PT-2)

Pricing Table 2: **Piping Erection and Painting**
Summary of Base Unit Rates

Unit no.	Description			
	Piping Erection and Painting		Currency:	

2.1000	**Handling**	**Unit**	**Base Unit Rate**	
			Small Bore	**Large Bore**
2.1101	Handling of (Prefabricated) Pipe Spools	m^1		
2.1102	Handling of Straight Run Pipe	m^1		
2.1200	Handling of Non-welded In-line Items	ea		
2.1300	Handling (for Certification) of Safety/Relief Valves	ea	See Pricing Table	

2.2000	**Joints**	**Unit**	**Base Unit Rate**	
			Small Bore	**Large Bore**
2.2100	Field Welds	ea		
2.2200	Flanged Joints	ea		
2.2250	Hydraulic Bolt Tensioning of Flanged Joints	ea		
2.2300	Screwed Joints	ea		

2.3000	**Bends**	**Unit**	**Base Unit Rate**	
			Small Bore	**Large Bore**
2.3000	Bends	ea		

2.4000	**Testing and Heat Treatment**	**Unit**	**Base Unit Rate**	
			Small Bore	**Large Bore**
2.4101	Radiographs (X-Ray)	ea		
2.4102	Radiographs (Gamma-Ray)	ea		
2.4110	Dye Penetrant Testing (DP)	ea		
2.4120	Magnetic Particle Testing (MP)	ea		
2.4140	Ultrasonic Testing (US)	ea		
2.4200	Post-Weld Heat Treatment (PWHT)	ea		
2.4300	Hydrostatic Testing, Flushing and Reinstatement	m^1		

2.5000	**Supports**	**Unit**	**Base Unit Rate**
2.5101	Prefabrication and installation of Small Bore Field Run Pipe Supports (<2 in.)	kg	
2.5102	Installation of Large Bore Pipe Supports (0–5 kg)	kg	
2.5103	Installation of Large Bore Pipe Supports (5–10 kg)	kg	
2.5104	Installation of Large Bore Pipe Supports (10–20 kg)	kg	
2.5105	Installation of Large Bore Pipe Supports (20–50 kg)	kg	
2.5106	Installation of Large Bore Pipe Supports (>50 kg)	kg	
2.5301	Spring Hangers (0–20 kg)	ea	
2.5302	Spring Hangers (20–50 kg)	ea	
2.5303	Spring Hangers (>50 kg)	ea	

2.6000	Field Painting	Unit	Base Unit Rate
2.6100	Cleaning and Painting of Welds	ea	
2.6101	Final Painting of Installed Pipe	m^1	
2.6201	Field Painting of Pipe Supports	kg	
2.6202	Field Painting of Pipe Supports (0–5 kg)	kg	
2.6203	Field Painting of Pipe Supports (5–10 kg)	kg	
2.6204	Field Painting of Pipe Supports (10–20 kg)	kg	
2.6205	Field Painting of Pipe Supports (20–50 kg)	kg	
2.6206	Field Painting of Pipe Supports (>50 kg)	kg	

2.8000	Modification Work	Unit	Base Unit Rate	
			Small Bore	Large Bore
2.8010	Tapering	ea		
2.8020	Drilling/Threading	ea		
2.8040	Threading	ea		
2.8101	Double Handling of Pipe Spools (for Modification Work)	m^1		
2.8102	Double Handling of Straight Run Pipe (for Modification Work)	m^1		
2.8103	Double Handling of In-line Items (for Modification Work)	ea		
2.8301	Cut (for Modification Work)	ea		
2.8302	Bevel (for Modification Work)	ea		
2.8400	Field Welds (for Modification Work)	ea		
2.8500	Flanged Joints (for Modification Work)	ea		
2.8550	Breaking Flanged Joints (for Modification Work)	ea		

Pricing Table 2: **Handling of Prefabricated Pipe Spools** (including Fittings)

| Material Type | Size | Schedule | Spool Length (m) | Base Unit Rate | Multipliers | | | Total Amounts |
					Material Type	Size	Schedule	
Small Bore								
Large Bore								
2.1101	Totals							

Pricing Table 2: **Handling of Straight Run Pipe**

Material Type	Size	Schedule	Pipe Length (m)	Base Unit Rate	Multipliers Material Type	Size	Schedule	Total Amounts
Large Bore								
2.1102	Totals							

Pricing Table 2: **Handling of Non-welded In-line Items**

| Size | Flange Rating | No. of Items | Base Unit Rate | Multipliers | | | Total Amounts |
				Size	Flange Rating	Special Item	
Small Bore							
Large Bore							
2.1200	**Totals**						

Pricing Table 2: **Handling** (for certification) **of Safety/Relief Valves**

Type of Valve	Inlet Size	Inlet Rating	Number of Valves	Unit Rate	Total Amounts
2.1300			Totals		

Pricing Table 2: **Field Welds**

Material	Size	Schedule	Special Item	No. of Welds	Base Unit Rate	Multipliers				Total Amounts
						Material Type	Size	Schedule	Special Item	
Small Bore										
Large Bore										
2.2100			Totals							

Pricing Table 2: **Flanged Joints**

| Size | Flange Rating | No. of Flanged Joints | Base Unit Rate | Multipliers | | Total Amounts |
				Size	Flange Rating	
Small Bore						
Large Bore						
2.2200	**Totals**					

Pricing Table 2: **Hydraulic Bolt Tensioning of Flanged Joints**

| Size | Flange Rating | No. of Flanged Joints | Base Unit Rate | Multipliers | | Total Amounts |
				Size	Flange Rating	
Small Bore						
Large Bore						
2.2250	Totals					

Pricing Table 2: **Screwed Joints**

| Material Type | Size | Schedule | No. of Screwed Joints | Base Unit Rate | Multipliers | | | Total Amounts |
					Material Type	Size	Schedule	
2.2300	Totals							

Pricing Table 2: **Bends**

Material Type	Size	Schedule	No. of Bends	Base Unit Rate	Multipliers			Total Amounts
					Material Type	Size	Schedule	
2.3000	**Totals**							

Pricing Table 2: **Radiographs** (X-Ray)

| Size | Schedule | No. of X-Rays | Base Unit Rate | Multipliers | | Total Amounts |
				Size	Schedule	
Small Bore						
Large Bore						
2.4101	**Totals**					

Pricing Table 2: **Radiographs** (Gamma-Ray)

Size	Schedule	No. of Gamma Rays	Base Unit Rate	Multipliers		Total Amounts
				Size	Schedule	
Small Bore						
Large Bore						
2.4102	Totals					

Pricing Table 2: **Dye Penetrant Testing** (DP)

Size	No. of DP Tests	Base Unit Rate	Multipliers	
			Size	**Total Amounts**
Small Bore				
Large Bore				
2.4110 Totals				

Pricing Table 2: **Magnetic Particle Testing** (MP)

Size	No. of MP Tests	Base Unit Rate	Multipliers	
			Size	Total Amounts
Small Bore				
Large Bore				
2.4120 Totals				

Pricing Table 2: **Ultrasonic Testing** (US)

Size	No. of US Tests	Base Unit Rate	Multipliers		Total Amounts
			Size		
Small Bore					
Large Bore					
2.4140 Totals					

Pricing Table 2: **Post-Weld Heat Treatment** (PWHT)

Size	Schedule	No. of Welds for Treatment	Base Unit Rate	Multipliers		Total Amounts
				Size	Schedule	
Small Bore						
Large Bore						
2.4200	Totals					

Pricing Table 2: **Hydrostatic Testing, Flushing and Reinstatement**

| Size | Flange Rating | Line Length (through all fittings) | Base Unit Rate | Multipliers | | Total Amounts |
				Size	Flange Rating	
Small Bore						
Large Bore						
2.4300	Totals					

Pricing Table 2: **Pipe Supports**

2.5100	SUPPORTS	Weight (kg)	Unit Rate	Total Amounts
2.5101	Prefabrication and Installation of Small Bore Field Run Pipe Supports (<2 in.)			
2.5102	Installation of Large Bore Pipe Supports (0–5 kg)			
2.5103	Installation of Large Bore Pipe Supports (5–10 kg)			
2.5104	Installation of Large Bore Pipe Supports (10–20 kg)			
2.5105	Installation of Large Bore Pipe Supports (20–50 kg)			
2.5106	Installation of Large Bore Pipe Supports (>50 kg)			
2.5100	Total Support Work			

2.5300	SUPPORTS	Each (ea)	Unit Rate	Total Amounts
2.5301	Spring Hangers (0–20 kg)			
2.5302	Spring Hangers (20–50 kg)			
2.5303	Spring Hangers (>50 kg)			
2.5300	Total Support Work			

Pricing Table 2: **Cleaning and Painting of Welds**

Paint System	Size	No. of Welds	Base Unit Rate	Multiplier Size	Total Amounts
☐			☐		
☐			☐		
☐			☐		
☐			☐		
2.6100	**Totals**				

Pricing Table 2: **Final Painting of Installed Pipe**

Paint System	Size	Equivalent Pipe Length (m)	Base Unit Rate	Multiplier Size	Total Amounts
2.6101	Totals				

Pricing Table 2: **Field Painting of Supports**

2.6200	PAINTING OF SUPPORTS	Weight (kg)	Base Unit Rate	Total Amounts
2.6201	Field Painting of Small Bore Pipe Supports			
2.6202	Field Painting of Pipe Supports (0–5 kg)			
2.6203	Field Painting of Pipe Supports (5–10 kg)			
2.6204	Field Painting of Pipe Supports (10–20 kg)			
2.6205	Field Painting of Pipe Supports (20–50 kg)			
2.6206	Field Painting of Pipe Supports (>50 kg)			
2.6200	**Subtotal Painting of Supports**			

Pricing Table 2: **Tapering**

Material Type	Size	No. of Taperings	Base Unit Rate	Multipliers Material Type	Size	Total Amounts
Small Bore						
Large Bore						
2.8010	Totals					

Pricing Table 2: **Drilling/Threading**

		Each	Base Unit Rate	Total Amounts
2.8020	**Drilling/Threading**			
	Totals			

Pricing Table 2: **Threading**

Size	No. of Threadings	Base Unit Rate	Multipliers Size	Total Amounts
2.8040 Totals				

Pricing Table 2: **Double Handling of Pipe Spools** (for modification work)

Material Type	Size	Schedule	Pipe Length (m)	Base Unit Rate	Multipliers			Total Amounts
					Material Type	Size	Schedule	
Small Bore								
Large Bore								
2.8101		Totals						

Pricing Table 2: **Double Handling of Straight Run Pipe** (for modification work)

Material Type	Size	Schedule	Pipe Length (m)	Base Unit Rate	Multipliers Material Type	Size	Schedule	Total Amounts
2.8102		Totals						

Pricing Table 2: **Double Handling of In-line Items** (for modification work)

| Size | Flange Rating | No. of Items | Base Unit Rate | Multipliers | | Total Amounts |
				Size	Flange Rating	
Small Bore						
Large Bore						
2.8103	Totals					

Pricing Table 2: **Cut** (for modification work)

Material Type	Size	Schedule	Number of Cuts	Base Unit Rate	Multipliers Material Type	Size	Schedule	Total Amounts
Small Bore								
Large Bore								
2.8301		Totals						

Pricing Table 2: **Bevel** (for modification work)

Material Type	Size	Schedule	Number of Bevels	Base Unit Rate	Multipliers			Total Amounts
					Material Type	Size	Schedule	
Small Bore								
Large Bore								
2.8302		Totals						

Pricing Table 2: **Field Welds** (for modification work)

Material Type	Size	Schedule	Number of Welds	Base Unit Rate	Multipliers			Total Amounts
					Material Type	Size	Schedule	
Small Bore								
Large Bore								
2.8400		Totals						

Pricing Table 2: **Flanged Joints** (for modification work)

| | | | | Multipliers | | |
Size	Flange Rating	No. of Flanged Joints	Base Unit Rate	Size	Flange Rating	Total Amounts
Small Bore						
Large Bore						
2.8500	Totals					

Pricing Table 2: **Breaking of Flanged Joints** (for modification work)

Size	Flange Rating	No. of Flanged Joints	Base Unit Rate	Multipliers Size	Flange Rating	Total Amounts
Small Bore						
Large Bore						
2.8550	Totals					

5. Examples

The purpose of this section is to give examples of piping pricing calculation according to the ECI Pricing System.

Three examples are given:

- Example 1 refers to Prefabrication of Large Bore Piping
- Example 2 refers to Erection of Large Bore Piping
- Example 3 refers to Fabrication and Installation of Small Bore Piping

To support these three examples, two isometrics have been drawn:

- Isometric nr 1007 rev 2 to support calculation examples nr 1 and nr 2
- Isometric nr 1003 rev 2 to support calculation example nr 3

The Base Unit Rates used in these three examples are derived from an example of what a contractual 'Summary of Reference Unit Rates' could be:

- Pricing Table 1: Piping Prefabrication and Painting
- Pricing Table 2: Piping Erection

Prices given in these tables are not a reflection of a pricing level in any particular currency.

They are just figures given for a better understanding of the ECI Pricing System.

5.1 Example of a Contractual Summary of Base Unit Rates

Pricing Table 1: **Piping Prefabrication and Painting**

Unit no.	Description		
	Piping Prefabrication & Painting	Currency: *Contractual*	

1.0000	Transportation	Unit	Base Unit Rate
1.0001	Transport Warehouse to Shop	ea	
1.0002	Transport Shop to Site	ea	

1.1000	Handling	Unit	Base Unit Rate
1.1102	Handling of Straight Run Pipe	m^1	

1.2000	Joints	Unit	Base Unit Rate
1.2100	Welds	ea	140.00

1.4000	Testing and Heat Treatment	Unit	Base Unit Rate
1.4101	Radiographs (X-Ray)	ea	
1.4102	Radiographs (Gamma-Ray)	ea	50.00
1.4110	Dye Penetrant Testing	ea	7.00
1.4120	Magnetic Particle Testing	ea	
1.4140	Ultrasonic Testing	ea	
1.4200	Post-Weld Heat Treatment	ea	
1.4300	Hydrostatic Testing and Flushing	ea	

1.6000	Painting	Unit	Base Unit Rate
1.6101	Painting of Pipe Spools (Paint System 1)	m^1	8.00
1.6102	Painting of Straight Run Pipe (Paint System 1)	m^1	
1.6202	Painting of Pipe Supports (0–5 kg)	kg	
1.6203	Painting of Pipe Supports (5–10 kg)	kg	
1.6204	Painting of Pipe Supports (10–20 kg)	kg	
1.6205	Painting of Pipe Supports (20–50 kg)	kg	
1.6206	Painting of Pipe Supports (>50 kg)	kg	

Pricing Table 2: **Piping Erection and Painting**

Unit no.	Description		
	Piping Erection and Painting	Currency: *Contractual*	

2.1000	Handling	Unit	Base Unit Rate	
			Small Bore	Large Bore
2.1101	Handling of Pipe Spools	m¹	35.00	100.00
2.1102	Handling of Straight Run Pipe	m¹		
2.1200	Handling of In-line Items	ea	60.00	170.00
2.1300	Handling of Safety/Relief Valves	ea	See Pricing Table	

2.2000	Joints	Unit	Base Unit Rate	
			Small Bore	Large Bore
2.2100	Field Welds	ea	45.00	190.00
2.2200	Flanged Joints	ea	45.00	100.00
2.2250	Hydraulic Bolt Tensioning of Flanged Joints	ea		
2.2300	Screwed Joints	ea	35.00	

2.3000	Bends	Unit	Base Unit Rate	
			Small Bore	Large Bore
2.3000	Bends	ea	52.00	

2.4000	Testing and Heat Treatment	Unit	Base Unit Rate	
			Small Bore	Large Bore
2.4101	Radiographs (X-Ray)	ea		
2.4102	Radiographs (Gamma-Ray)	ea		70.00
2.4110	Dye Penetrant Testing	ea		
2.4120	Magnetic Particle Testing	ea		
2.4140	Ultrasonic Testing	ea		
2.4200	Post-Weld Heat Treatment	ea		
2.4300	Hydrostatic Testing	m¹	10.00	20.00

5.2 Example 1: Prefabrication of Large Bore Piping (Isometric nr 1007 rev 2)

This calculation is based on the Base Unit Rates shown in **Pricing Table 1: Piping Prefabrication and Painting** and in **Pricing Table 2: Piping Erection for installation** and primer coating of small bore attachments such as o'lets

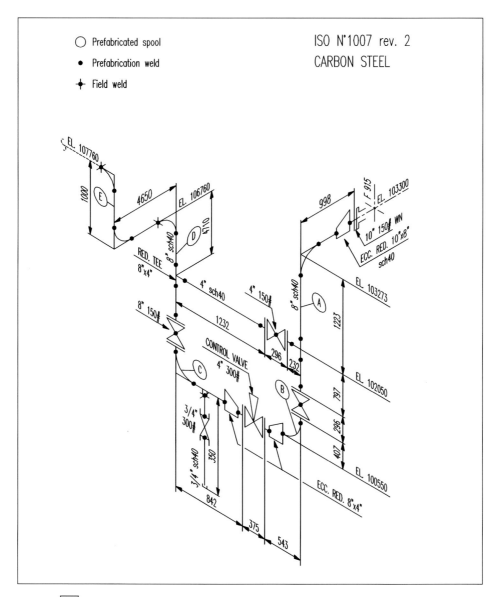

PRICING CALCULATION EXAMPLE nr 1:
Prefabrication of Large Bore Piping

Isometric nr 1007 rev 2
Carbon Steel

| Prefabrication Welds | | | | | | Multipliers | | | | Total Amounts |
Material Type	Size (in.)	Schedule	Special Items	No. of Welds	Base Unit Rate	Material Type	Size	Schedule	Special Items	contractual currency
CS	4	sch 40		5	140.00	1.00	0.80	1.00		560.00
CS	8	sch 40		18	140.00	1.00	1.40	1.00		3,528.00
CS	10	sch 40		1	140.00	1.00	1.80	1.00		252.00
CS	3/4		outlet 90°	1	45.00	1.00	1.00		4.00	180.00

| NDE: Prefabrication Radiographs (Gamma-Ray) | | | | | | Multipliers | | | | |
Size (in.)	Schedule			No. of Gamma Rays	Base Unit Rate	Size	Schedule			
8	sch 40			2	50.00	1.10	1.00			110.00

| NDE: Prefabrication Dye Penetrant Testing | | | | | | Multipliers | | | |
Size (in.)				No. of DP Tests	Base Unit Rate	Size			
8				2	7.00	1.20			16.80
4				1	7.00	0.90			6.30

| Painting (shop) of Prefabricated Pipe Spools | | | | | | Multipliers | | | |
Paint System	Size (in.)			Equivalent Length (m)	Base Unit Rate	Size			
P1	4			2	8.00	0.70			11.20
P1	8			20	8.00	1.30			208.00
P1	10			0.5	8.00	1.60			6.40
P1	3/4			0.15	8.00	0.30			0.36

GRAND TOTAL *(contractual currency)* **4,879.06**

5.3 Example 2: Erection of Large Bore Piping (Isometric nr 1007 rev 2)

This calculation is based on the Base Unit Rates shown in **Pricing Table 2: Piping Erection**

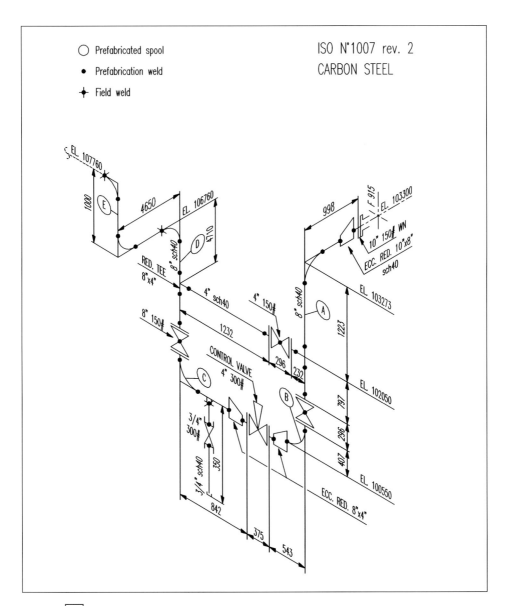

PRICING CALCULATION EXAMPLE nr 2:
Erection of Large Bore Piping

Isometric nr 1007 rev 2
Carbon Steel

Handling of Large Bore Pipe Spools				Spool Length (m)	Base Unit Rate	Multipliers				Total Amounts (contractual currency)
Material Type	Size (in.)	Schedule				Material Type	Size	Schedule		
CS	4	sch 40		1.64	100.00	1.00	0.90	1.00		147.60
CS	8	sch 40		15.92	100.00	1.00	1.10	1.00		1,751.20
CS	10	sch 40		0.28	100.00	1.00	1.30	1.00		36.40

Handling of Large Bore In-line Items						Multipliers				
Material Type	Size (in.)	Flange Rating (lbs)	Special Item	No. of Items	Base Unit Rate	Material Type	Size	Flange Rating	Special Item	
CS	4	150		1	170.00	1.00	0.80	1.00	1.00	136.00
CS	4	300	CV	1	170.00	1.00	0.80	1.15	2.00	312.80
CS	8	150		2	170.00	1.00	1.25	1.00	1.00	425.00

Large Bore Field Welds						Multipliers				
Material Type	Size (in.)	Schedule	Special Item	No. of Welds	Base Unit Rate	Material Type	Size	Schedule	Special Item	
CS	8	sch 40		2	190.00	1.00	1.40	1.00	1.00	532.00

Large Bore Flanged Joints						Multipliers				
Size (in.)	Flange Rating (lbs)			No. of Flanged Joints	Base Unit Rate	Size	Flange Rating			
4	150			2.00	100.00	0.80	1.00			160.00
4	300			2.00	100.00	0.80	1.15			184.00
8	150			4.00	100.00	1.25	1.00			500.00
10	150			1.00	100.00	1.50	1.00			150.00

Handling of Small Bore Piping						Multipliers				
Material Type	Size (in.)	Schedule		Line Length (m)	Base Unit Rate	Material Type	Size	Schedule		
CS	3/4	40		0.24	35.00	1.00	1.00	0.90		7.56

Small Bore Piping Welds						Multipliers				
Material Type	Size (in.)	Schedule	Special Item	No. of Welds	Base Unit Rate	Material Type	Size	Schedule	Special Item	
CS	3/4	40	SW	3	45.00	1.00	1.00	0.90	0.60	72.90

Screwed Joints						Multipliers				
Material Type	Size (in.)	Schedule		No. of Screwed Joints	Base Unit Rate	Material Type	Size	Schedule		
CS	3/4	40		1.00	35.00	1.00	1.00	1.00		35.00

NDE: Erection Radiographs (Gamma-Ray)						Multipliers				
Size (in.)	Schedule			No. of Gamma Rays	Base Unit Rate	Size	Schedule			
8	40			1	70.00	1.15	1.00			80.50

Hydrostatic Testing						Multipliers				
Size (in.)	Flange Rating (lbs)			Line Length (m)	Base Unit Rate	Size	Flange Rating			
3/4	300			0.35	10.00	1.00	1.00			3.50
4	150			2.31	20.00	0.90	1.00			41.53
8	150			16.51	20.00	1.10	1.00			363.26
10	150			0.28	20.00	1.30	1.00			7.28

GRAND TOTAL (contractual currency) **4,946.60**

5.4 Example 3: Fabrication and Installation of Small Bore Piping (Isometric nr 1003 rev 2)

This calculation is based on the Base Unit Rates shown in **Pricing Table 2: Piping Erection**

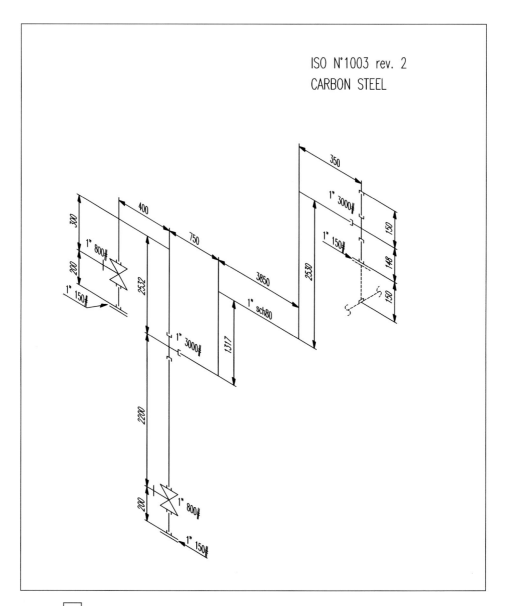

ISO N°1003 rev. 2
CARBON STEEL

PRICING CALCULATION EXAMPLE nr 3:
Fabrication and Installation of Small Bore Piping

Isometric nr 1003 rev 2
Carbon Steel

Handling of Small Bore Piping						Multipliers				Total Amounts
Material Type	Size (in.)	Schedule		Line Length (m)	Base Unit Rate	Material Type	Size	Schedule		(contractual currency)
CS	1	sch 80		14.93	35.00	1.00	1.10	1.00		574.81

Handling of Small Bore Non-Welded In-line Items						Multipliers				
Material Type	Size (in.)	Rating (lbs)	Special Item	No. of Items	Base Unit Rate	Material Type	Size	Rating	Special Item	
CS	1	900		2.00	60.00	1.00	1.10	1.50	1.00	198.00

Flanged Joints						Multipliers				
Size (in.)	Flange Rating (lbs)			No. of Flanged Joints	Base Unit Rate	Size	Flange Rating			
1	150			3.00	45.00	1.10	1.00			148.50

Screwed Joints						Multipliers				
Material Type	Size (in.)	Schedule		No. of Screwed Joints	Base Unit Rate	Material Type	Size	Schedule		
CS	1	sch 80		14.00	35.00	1.00	1.10	1.00		539.00

Bends						Multipliers				
Material Type	Size (in.)	Schedule		No. of Bends	Base Unit Rate	Material Type	Size	Schedule		
CS	1	sch 80		6.00	52.00	1.00	1.10	1.00		343.20

Hydrostatic Testing						Multipliers				
Size (in.)	Flange Rating (lbs)			Line Length (m)	Base Unit Rate	Size	Flange Rating			
1	150			14.93	10.00	1.10	1.00			164.23

GRAND TOTAL (contractual currency) **1,967.74**